The Theory of

Electrogravity

Seyed Sadegh Mousavi

Translated by:

Mozhghan Nasiri

Seyed Sadegh Mousavi / The Theory of Electrogravity

ISBN: 978-1-3999-7803-3

(2023)

4

To professor Yousef Sobouti

It is always a source of pleasure when a great and
beautiful idea proves to be correct in actual fact.

Albert Einstein [letter to Sigmund Freud]

Contents:

8

1- Introduction:

The contents of this book, which are about the Electrogravity theory, have emerged from more than twenty years of my intensive effort and research on unifying gravitation with electromagnetism.

Before reading this book, for better understanding, one should be familiar with the tensor calculations and general relativity theory. To make it easy to understand, first I will explain the mathematical background which is necessary to realize it and then make a glimpse to the general relativity theory. By studying the existing theories in this field, it becomes evident that, although these theories have been changed several times during the history of physics and became much more complete than before, and they have solved many scientific problems, but, there are some other questions yet, that they are unable to answer them.

I firmly believe that, it is the duty of all of us, to try and take science one step forward and contribute towards advancing science, which I believe that, the progress in science ultimately enhances the quality of human's life.

Until now, many physicists have attempted to enter the gravitation in electromagnetism and Quantum mechanics, for example by proposing the graviton particle for gravitation or other ways, but they haven't been successful.

However, we pose a question, why we don't try to enter the quantum in geometry and in general relativity. Maybe this way results better. Of course it will be a hard task, but I think it is possible.

Unfortunately, some researchers in physics hold the misconception that, the electromagnetism exist in general relativity and it is adequately addressed within the theory of general relativity, and we don't need to waste our time for entering electromagnetism in the gravity field equations. But it is wrong, and this flawed understanding not only deters them from tackling this unresolved issue of Electrogravity but also discourages others from exploring this crucial matter. When I spoke with some of them who claimed this, and asked their reason for this claim, surprisingly, I encountered physicists who subscribed to this viewpoint, citing examples of exact solutions of Einstein's field equations, and they said, for example electromagnetism has been entered in Rreissner-Nordstrom Solution of the Einstein field equation. But the second group pointed to the Kaluza[1] theory.

I think, at the beginning of this book, it is necessary to answer the first group that, firstly, Einstein, himself, had seen the Reissner-Nordström solution, so why did he try again to enter the electromagnetism in his field equations? Secondly, we must pay attention to the issue that, by importing an electric charge into a solution of the Einstein field equations, although we accept in mathematics that, the value of the electric charge which is considered as a constant value in the metric, gives a correct answer for a solution of Einstein field equations, but it does not provide a comprehensive explanation of the geometrical behaviors of the electric charge unit and electromagnetic field. Therefore, we should try to enter the electric charge unit and electromagnetism in general relativity field

---

[1] T. Kaluza, Sitzungsber. Preuss. Akad. Wiss, 966 , 1921.

equations with a more logical method, not to enter the electromagnetism just in the solutions of the field equations alone.

But about the second question, although Kaluza tried to enter the electromagnetism in general relativity , but eventually his theory refused to be accepted by scientific society.

So, as it is mentioned above, since 1914, that Albert Einstein proposed a new and complete definition of gravitation, there have been several attempts to unify electromagnetism and gravitation, such as Kaluza theory and some other attempts in the area of Quantum mechanics, all of which were unsuccessful in unifying electromagnetism with gravitation. For example, several years after the suggestion of Kaluza theory, in 1939 Einstein rejected his theory in a letter mentioning that: "*As no arbitrary constants occur in the equations, the theory would lead to electromagnetic and gravitation fields of the same order of magnitude. Therefore one would be unable to explain the empirical fact that the electrostatic force between two particles is so much stronger than the gravitational force. This means that a consistent theory of matter could not be based on these equations.*" [1]

Einstein, himself, tried a lot to do this unification, as he commented: "*It would be a great step forward to unify in a single picture the gravitational and electromagnetic fields, then there would be a worthy completion of the epoch of theoretical physics …*" [2].

---

[1] A. Einstein. Letter to V. Bargmann, July 9. 1939, AE 6-207
[2] A. Einstein. Äther und Relativitätstheorie. Springer,1920

12

Einstein was trying to unify these two fields in a theory based on fields and not particles, as it has been mentioned in his paper in 1935: "*A complete field theory knows only fields and not the concepts of particles and motions*"[1]. Einstein wanted the fields to be absorbed in geometry and he wanted to formulate electromagnetism in geometry, same as gravitation.

But electromagnetism has not been absorbed in geometry in any of the previous theories.

Einstein, in his field equations, showed that, the effect of gravitational field on a space-time is explained with a symmetric rank 2 tensor, namely Ricci tensor. Also, it is known that, the effect of electromagnetic filed on a space-time is explained with a rank 2 antisymmetric tensor (electromagnetic tensor), which satisfies Maxwell's equations.

In the real world of physics, both electromagnetic and gravitational fields exist in a space-time simultaneously. Therefore, the space-time must be simultaneously formulated using two second rank tensors, one of which is symmetric and the other antisymmetric. But what is the relationship between these two tensors and how can they be introduced in a single equation? In this book, a new theory is proposed for unifying gravitation with electromagnetism in a five dimensional space-time, while considering the above mentioned tensors. But Before delving deeper, let's briefly revisit some noteworthy previous theories.

2- Different ideas about electromagnetism and gravitation

[1] A. Einstein and N. Rosen. The particle problem in the General theory of relativity. Physical Review,48(1):73-77,1935.

The electric charge unit is familiar to all of us. The journey to understand the nature of the electric charge unit has involved various stages throughout the history of physics.

First in 1897, Josef john Thomson introduced the particle theory of the electron, postulating that electrons are particles which are 1000 times smaller than atoms. He was awarded the Nobel Prize for this discovery. Several years later, his son, George Paget Thomson, provided evidences that electrons are not particles but rather exhibit wave-like properties. Again they gave the Nobel prize to him for this discovery. So until here, one won the Nobel prize for proving that , the electron is particle, and some years later the other one won the Noble prize for proving that the electron is not particle, and it is wave. Several years later, Louis de Broglie, proposed the wave-particle duality of electrons, and with this theory he won the Nobel prize in physics in 1929. Essentially, he demonstrated that an electron can exhibit particle and wave behaviors simultaneously.

The Davison Germer experiment[1] was a physical experiment performed by Clinton Davison and Lester Germer between 1923 and 1927 that confirmed the wave-particle duality hypothesis introduced by Louis de Broglie in 1924 and was a turning point in the development of quantum mechanics.

---

[1] *Davisson, C. J.; Germer, L. H. (1928). "Reflection of Electrons by a Crystal of Nickel". Proceedings of the National Academy of Sciences of the United States of America. 14 (4): 317–322*

14

Finally, with the discovery of quantum science, it was accepted that, an electron is both particle and wave, and depending on the experimental method used for measurement , it manifests itself as a particle or wave.

Subsequently, another theory emerged, suggesting that the electron is none of these, but electron is string and ..... .

Now we say, wait a moment. Let's pause and think a bit more. May be the nature of the electric charge unit is something else, that revealing itself in different forms under various conditions.

Here the important question is, why everyone always underestimates an electron? and in particle theory of electron, electrons are considered to be particles with very small mass (10^{-31} kg) to the point where many regard its mass as virtually zero.

In classical particle theory of electron, if we think about this issue a little positively and analyze this from a different perspective, we find that the mass of this particle is disproportionately large compared to its size. In fact, it's among the particles in nature with the highest mass-to-size ratio. The disparity lies in the fact that, this mass is seemingly small for a very small particle, but since it has very small size, it is remarkably substantial in proportion to its size. Don't you believe? Let's examine the mass density of an electron in classical particle theory of electron:

$M_e = 9.109*10^{-31}$ (kg), and $R_e = 2/82*10^{\wedge}(-15)$ (m), and $V_e = 9.38*10^{-44}$ (m^3) (2-1) where V_e is the volume of an electron , then the mass density of an

electron is: $\rho_e = 9.7*10^{12}$ ($\frac{kg}{m^3}$) and comparatively we know the mass

density of gold metal is: $\rho_{gold} = 19320$ ($\frac{kg}{m^3}$)

So as it is seen, if the size of this particle was one cubic meter, its mass would be 9700 billion (kg). In the macroscopic world, even the densest metals, such as gold, appear as mere specks when compared to this hypothetical massive object. If we had such a heavy object in a macroscopic scales, according to Einstein's theory of general relativity, we would expect it to strongly curve the space time around it. So, the first thing that comes to the mind is: maybe in microscopic scale, this particle with such a large density also creates this extreme curvature around itself.

So it must be contemplated that, may be, either there is a problem in the old theories mentioned above, or there is a problem in the above numerical calculations!

Here It is important to emphasize that, in microscopic scales, nature's laws introduce constraints on these small particles, setting them different from the macroscopic scales, such as the presence of electric charge in these particles and the property of repulsion of two similar charges which restrict them from gravitational attraction. These things show that, the Einstein's equations of general relativity alone, unable to describe the complete behaviors of these particles.

Therefore, we need a more complete equation than Einstein's equations, that it can simultaneously explain the gravitational geometric behaviors and the electromagnetic geometric behaviors of these microscopic charged particles.

For more consideration and resolve the above contradictions, first we briefly review the most important previous theories related to gravity and electromagnetism, then try to find a more complete and better theory.

At this juncture, it's essential to express my profound respect for figures like Newton, Einstein and …., who persevered through many challenges in their life and contributed to the advancement of science. They are truly great human beings. To pay homage to them, first I will review a summary of their theory and then move on to express the theory of Electrogravity. While I hold the same level of respect for figures like Faraday, Ampere, Thomsons, Maxwell, and others who explored the electromagnetic properties of electric charges and other properties of the electromagnetic fields, to maintain the book's conciseness, their theories will not be explored within this book.

3- Newton's Gravity:

Since the time of the Greeks, two fundamental phenomena had intrigued thinkers: the attraction of objects towards the Earth when released and the motion of celestial bodies, including the Sun and the Moon, which were regarded as planets in that era. Historically, these two phenomena were considered separately. The first one, who presented a scientific explanation of the gravity force in his Principles book, was Isaac Newton. A remarkable achievement of Newton was his revelation that these seemingly distinct phenomena were, in fact, interconnected and governed by the same laws. This insight marked a significant turning point in our understanding of the universe.

In the year 1665, after the school closed due to the prevalence of plague, Newton who was 23 years old at the time went from Cambridge to Lincoln- -shire.

It was approximately fifty years later when he reflected on that year and stated: "In that same year (1665), it dawned on me that the force necessary to keep the Moon in its orbit and the force of gravity acting on the Earth's surface were, to a good approximation, similar."

William Stokely, one of Isaac Newton's young friends, writes that when he was having tea with Isaac Newton under the apple trees of an orchard, Isaac Newton told him that the idea of gravity came to his mind in such a place. Stokely writes: While he was sitting and thinking, the falling of an apple catches his attention and he realizes the concept of gravity. After that, he gradually applies the property of gravity to the movement of the earth and celestial bodies and …. .

Remarkably, Newton didn't publish the results of his calculations until 1687, nearly 22 years after he had grasped the fundamental concept of gravity. Two significant reasons contributed to this delay: one was, he didn't know the Earth's radius, which was essential for his calculations, and the other was Newton's general reluctance to publish his findings. Known for his introverted and shy nature, he abhorred debate and controversy. It was Edmund Halley, the namesake of Halley's Comet, who ultimately persuaded Newton to publish his monumental work, "Principles."

Within the pages of "Principles," Newton's theory of gravity expanded far beyond the anecdotal apple's interaction with Earth. He applied his law of gravity universally, extending its influence to all objects. With this unifying concept, Newton successfully reconciled terrestrial and celestial mechanics, previously regarded as separate disciplines. His remarkable feat was to merge them into a single, comprehensive theory.

In fact, Newton played a unique role in serving the science of the 17th century. Isaac Newton's pioneering work enabled the explanation of planetary motions in our solar system and the behavior of falling objects near Earth's surface under a single, unifying concept. In doing so, he bridged the gap between terrestrial and celestial mechanics, setting the stage for a new scientific era. In 1687, he finally published his seminal work, "Principles."

4-Newton's law of universal gravitation:

The gravitational force between two particles with masses m1 and m2, separated by a distance r, is governed by Isaac Newton's Law of Universal Gravitation. The magnitude of this force is determined by the equation:

$$F = \frac{Gm_1m_2}{r^2}$$

(4-1)

Where G is a universal constant and its value is the same for all pairs of particles. This is Isaac Newton's law of universal gravitation. To comprehend this law more thoroughly, let's highlight some of its key features.

Gravitational forces between two particles are a pair of action-reaction forces (action and reaction). The first particle exerts a force on the second particle whose direction is towards the first particle (gravity) and along the line that connects the two particles. Conversely, the second particle exerts an equal but opposite force on the first particle, directed towards itself (gravity), and also along the connecting line. While these forces share the same magnitude, their directions oppose each other.

The universal constant G is a scalar quantity, while g is a vector quantity with dimension $\frac{L}{T^2}$, which is neither global nor constant, as it varies at different locations on Earth due to changes in the distance to the Earth's center.

The value of G can be obtained by performing detailed experiments. This work was done for the first time by Lord Cavendish in 1798. Presently, the accepted value for G is: $G = 6.67 \times 10^{-11}$

When an apple falls and accelerates towards the ground, simultaneously, the Earth experiences an upward acceleration towards the apple. However, given the Earth's immensely greater mass, the gravitational force exerted by the apple is negligible in comparison. Newton's gravitational equation can be employed to elucidate the Earth's orbit around the Sun.

The gravitational formula based on this equation substantiates Galileo's assertion that it is the Earth that orbits the Sun. Indeed, the gravitational force formula allows for the prediction that moons and planets follow elliptical orbits, aligning with Kepler's findings following his analysis of Tycho Brahe's observations.

For over two centuries, Newton's Law of Gravitation governed the field of astronomy, and it seemed to address the gravitational conundrum entirely. This formula was utilized to interpret a broad array of celestial phenomena. Nonetheless, Newton himself remained uncertain about the completeness of his comprehension of the natural world. He expressed his doubt in these words: "I do not know what I may appear to the world, but to myself I seem to have been only like a boy playing on the seashore, and diverting myself in now and then finding a smoother pebble or a prettier shell than ordinary, whilst the great ocean of truth lay all undiscovered before me".

It was Albert Einstein who first realized that there was more to gravity than Newton thought. After writing his treatise in 1905, he devoted all his attention and concentration to the development and evolution of the special relativity theory to the theory of general relativity. Through groundbreaking insights and a fundamentally different perspective, he redefined our understanding of gravitational forces between celestial bodies, including planets and moons.

5-Faraday's efforts on unifying electromagnetism and gravitation

Michael Faraday was one of the pioneers in seeking connection between electricity and gravity. He was interested in the fundamental forces of nature and spent a considerable amount of his time exploring the possibility of a connection between gravity and electromagnetism. In 1840, He embarked on a series of experiments to delve into the potential interplay between gravity and electromagnetism. His aim was to uncover a connection between these two distinct fields, but his experiments didn't yield the results he had anticipated. He found no direct evidence of a relationship between gravity and electromagnetism.

Despite these outcomes, Faraday remained steadfast in his belief that there had to be some underlying connections between these two realms. He held the conviction that a fundamental principle governed all physical phenomena, encompassing both gravity and electromagnetism. In 1843, Faraday published a paper titled "On the Possible Relation of Gravity to Electricity," documenting his diligent experiments seeking a connection between gravitation and electromagnetism. Regrettably, as mentioned earlier, his investigations failed to reveal any significant effect. He concluded his work with these words, "Here end my trials for the present. The results are negative, but they do not shake my strong feeling of an existence of a relation between gravity and electricity…"

While Faraday's experiments didn't deliver a breakthrough in our understanding of the relationship between gravity and electromagnetism, they constituted a pivotal step in the evolution of modern physics. His endeavors laid the foundation for future generations of scientists to explore the fundamental forces of nature and craft new theories to unify them. In 1865, James Clerk Maxwell attempted this unification in his paper, "A Dynamical Theory of the Electromagnetic Field," where he even formulated some equations in pursuit of this goal. Later, towards the end of the 19th century, Oliver Heaviside drew an analogy between gravitation and electromagnetism.

Many scientists, including Lorentz, Poincaré, and Minkowski, had postulated the possibility of gravitational fields in the early years of special relativity. Nevertheless, their endeavors proved fruitless, because Albert Einstein later redefined the gravitation in his General Theory of Relativity, altering the definition they had employed in their research (which had been based on Newton's definition). In general relativity, gravity isn't perceived as a force; instead, it is depicted as an outcome of the curvature of space-time. This revolutionary theory reshaped our understanding of gravity and the cosmos.

6- Einstein's gravity and General Relativity:

After years of relentless effort, finally Albert Einstein succeeded in conveying some facets of his general theory of Relativity to the scientific community of his time, which offered a much more comprehensive explanation of gravity.

He was very annoyed in this way and at first the scientific community refused to accept his theory.

Although after passing years, they knew what a valuable theory he gave, but due to the lack of comprehension of his theory by many physicists of that time, they did not dare to give Einstein's Nobel Prize to his theory of general relativity, because they didn't understand it's concepts, and later they gave the Nobel prize for a relatively simple discovery of Einstein's photoelectric effect. It was simple in comparison to his general relativity theory and really it is not comparable to his theory of general relativity at all.

To gain a better grasp of the theory of Electrogravity, one must has a good understanding of the General Relativity, so at first, let's begin by providing a simple introduction to the concept of general relativity and the notion of curved space-time. We compare the fundamental distinction between Einstein's view of the universe and Newton's perspective. According to Newton, mass instructs gravity how to exert force, and that force guides mass on how to move. Einstein, on the other hand, introduced a groundbreaking idea: mass instructs space-time how to bend, and this bending dictates the motion of mass.

When we discuss space-time curvature, we refer to a four-dimensional construct, encompassing three spatial dimensions and one dimension of time. This conceptual framework arises from Einstein's theory of Special Relativity, which uncovered the interconnectedness of space and time. At high speeds, objects exhibit phenomena like Lorentz contraction (shortening in space) and time dilation (expansion of time).

Einstein's perception of gravity diverged significantly from Newton's. While Newton's theory described gravity as a force exerted between masses, Einstein proposed that massive objects warp the space-time around them. Objects in free fall follow the path of least resistance in this curved space-time, resulting in what appears as a gravitational force. But what does it mean when we talk about space-time curvature? Space-time in general relativity has four dimensions, namely three spatial dimensions and one time dimension. To visualize the issue, consider a two-dimensional surface. If the surface is two-dimensional (such as the surface of a table), the shortest distance between two points is a straight line. Parallel lines never intersect on a surface. If the two-dimensional surface is curved, the shortest distance between two points becomes a curved line. On a spherical surface, the shortest distance is the great circle. This curvature leads to parallel lines converging or diverging. Just as we have flat or curved two-dimensional surfaces, flat volumes or a four-dimensional curve is also possible.

The curvature can be comprehended, but what is its connection with gravity? According to Newton's thinking, it has nothing to do with curvature, and a heavy object exerts force on other objects. But Newton's view can't answer the question of what carries force in an empty space. Einstein says that a heavy body warps the space-time around it. Objects in free fall follow the shortest path between two points, and therefore, they travel curved lines in the curved space-time. Einstein also says that photons also travel curved paths in curved space-time, even though they have no mass.

To illustrate Einstein's theory of gravity, imagine a rubber sheet stretched over a frame. This sheet is perfectly flat. If you place a small marble on it, it will follow a straight line without deflection. Now, introduce a heavy ball onto the sheet on the screen. This ball is so heavy which causes the plate to dent around it. Now drop the same marble on the plate and around the heavy ball. The curvature of the rubber plate causes the marble to follow a curved path. An observer who cannot see the curvature of the rubber surface from a distance will say that, the heavy ball is exerting a force on the marble and has deflected it from the right path.

Therefore, when an object travels the shortest path on a curved surface, it seems that a force is being applied to it. Similarly, when an object travels the shortest path in a four-dimensional space-time, it seems that a force (gravitational force) is applied to it.

7-Curved Surfaces

One of the most remarkable aspects of general relativity is its foundational reliance on the concept of curved space-time. While everyone intuitively knows what a curved surface is, or rather what it looks like in two dimensions, but most people get confused when it comes to extending the concept to three or more dimensions. This difficulty arises from the inability to visualize a four-dimensional space in which a three-dimensional space appears curved. Therefore, we first need to establish the intrinsic meaning of surface curvature, without reference to the surrounding space. These intrinsic properties are characteristics that solely depend on the size ratios on the surface and remain unaltered when the surface is bent without stretching or tearing. For example, a flat sheet of paper and another sheet that is bent roughly cylindrical or roughly bent into a cone are locally equivalent to each other. However, when the cylinder or cone is closed, these surfaces are still "locally" equivalent but not overall. Intrinsic properties are those that remain unchanged when a surface is bent without stretching or tearing. An essential intrinsic property of a surface can be seen in its set of geodesic lines. These are paths along which the distance between two nearby points is minimized. Geodesics are intrinsic because they only depend on the size of the distances in the surface, that is, if the surface is bent, it will still remain a geodesic. According to Fermat's principle, it can be assumed that light travels along geodesics in this world. Therefore, the world appears to every observer to be flat (flat) and is indistinguishable from a plane tangent to it. However, it is crucial for the observer not just to observe but to measure these properties.

Using elementary geometry and Taylor's series for sine, we have:

$$\eta = \theta\left(a\sin\frac{r}{a}\right) = \theta\left(r - \frac{r^3}{6a^2} + ...\right) = \theta\left(r - \frac{1}{6}Kr^3 + ...\right) \tag{7-1}$$

Which lead to:

$$C = 2\pi\left(r - \frac{1}{6}Kr^3 + ...\right), \quad A = \pi\left(r^2 - \frac{1}{12}Kr^4 + ...\right)$$

(7-2)

to get A. $A = \int C dr$ here we have used the relation

These expansions yield the following two different relations for the curvature K:

$$K = \frac{3}{\pi}\lim_{r\to 0}\frac{2\pi r - C}{r^3} = \frac{12}{\pi}\lim_{r\to 0}\frac{\pi r^2 - A}{r^4} \tag{7-3}$$

It can be shown that the validity of equation (7-1) extends up to the specified order for any sufficiently differentiable surface. In other words, the expansion of neighboring geodesics that are drawn in different directions from point P, up to the third order of r, follows the pattern described in (7-1), where K is a number which is unique to point P. This number is called the curvature (Gaussian) of the surface at the point P. Therefore, equations (7-2) and (7-3) are used quite generally.

Furthermore, it can be shown that, in every surface with K=0 throughout, the surface is inherently flat like a plane. While every surface with K=1/a² at every point, is inherently a sphere with radius a.

8- Curved spaces with more dimensions

The concepts related to the intrinsic geometry of surfaces can be expanded to spaces with more dimensions such as three-dimensional space. In particular, geodesics are defined in each dimension just like in the two-dimensional case.

Two significant theorems can be established for sufficiently "well-behaved" spaces:

(a) For any given point and direction, there exists a unique geodesic, and
(b) Within a sufficiently small neighborhood of point P, any other point can be connected to point P by a unique geodesic route.

By creating geodesic spheres with radius r (rather than circles) and comparing their surface or volume to Euclidean values, we can generalize the concept of curvature from two dimensions to three dimensions. It is logically quite conceivable that precise measurements of this kind might reveal the deviation of our space from flat space. Gauss, himself carried out several experiments to determine the curvature of space, but nothing could be directly detected with the instruments available, that time.

Describing the curvature properties of spaces with more than two dimensions requires more than one number. By considering branched geodesics originating from point P in stretches defined by two directions, p and q, we can speak of these geodesics generating a transitive geodesic plane at point P. The curvature of this plane at point P, denoted as 'k',characterizes the space's curvature at P in the direction of (p, q). This notation extends to six directions in three dimensions and 20 directions in four dimensions. The geodesic plane in curved space is analogous to a plane in flat space, but it typically has distinctive properties at a single point.

If light travels along the geodesics, an observer at point P on each generator will see a straight line, and thus the entire geodesic plane from that point will appear to be a plane.

When the curvature at point P, denoted as K, is independent of direction, we label point P as isotropic. In this case, all the information related to curvature lies in the identification of the area S or the volume V of a small geodesic sphere of radius r. It is easy to see that in three dimensions, S and V are given by the following formulas:

$$S = 4\pi \left(r^2 - \frac{1}{3}kr^4 + ... \right) , \quad V = \frac{4}{3}\pi \left(r^3 - \frac{1}{5}kr^5 + ... \right) \qquad (8\text{-}1)$$

The first equation is obtained from (7-1): the ratio of S to the Euclidean value $4\pi r^2$ must be equal to the square of the η to the Euclidean value θr. Then the second formula is obtained from the $V = \int S dr$. If all points in a space are isotropic, this implies that the curvature is uniform across all points, this is a principle known as Schor's theorem. In this case, the space is said to possess constant curvature. Schor's theorem can be briefly stated as , isotropy of all points implies homogeneity.

Now let's discuss the concept of a three-dimensional equivalent of a sphere, characterized by a constant and positive curvature for example $1/a^2$. By employing the equation (7-1), we can derive the exact form of the equation (8-1) for such a sphere.

$$S = 4\pi a^2 \sin^2 \frac{r}{a} \ , \quad V = 2\pi a^2 \left(r - \frac{a}{2} \sin \frac{2r}{a} \right) \tag{8-1}$$

We consider a normal two-dimensional sphere with a curvature k. Let's see what happens if we draw circles around a given point. With the increase of the geodesic radius r, the circumference of the circles increases at first, but after reaching a maximum, it decreases again and finally reaches zero, for $r = \pi a$. If we lived in a 3D curved sphere with curvature $1/a^2$ and drew concentric geodesic spheres around us, we would see that their area first increases as the geodesic radius r increases (but not as fast as the Euclidean mode) until it reaches a maximum $4\pi a^2$ per radius $1/2\pi a$, which includes the volume $\pi^2 a^2$ (compare with (8-1)). After that successive spheres contract and finally we reach a sphere with a radius $r = \pi a$ that has zero surface area and yet includes all our space, essentially, its surface becomes a single point. While the volume of this three-dimensional sphere is finite and equal to a certain value, it lacks boundaries and a central point – every point within it is equivalent to every other point. Each geodesic plane within this space forms a two-dimensional sphere. Just like a great circle divides a two-dimensional sphere into two equal halves, these geodesic planes similarly bisect the space. To illustrate the concept of geodesic deviation, which relates to what was discussed earlier, if we take the derivative of the third equation in (7-1) with respect to 'r' twice, we will get the following with the first order approximation:

$$\ddot{\eta} = -K\eta \left(0 \equiv d/dr\right)$$
(8-2)

Therefore, the second derivative of two geodesics that intersect at a small angle can be used as a very direct measure of the curvature of a surface. Equation (8-2) is actually true for any two adjacent geodesics, even if they do not intersect within the desired range.

9- Riemannian spaces

Usually the curved spaces discussed are actually Riemannian. We cannot define Cartesian coordinates on a curved surface in the same way we do for flat surfaces, where a grid of complete squares serves as coordinates, because if we could, naturally, we had a surface. However, some surfaces possess "natural" coordinates due to their symmetry properties. For instance, when dealing with a sphere, it's common to employ width (x) and length (y) as coordinates to specify points on the sphere. In general, two arbitrary coordinate lines can be drawn on a surface, labeled with x=...,-2,-1,0,1,2,... and y=...,-2,-1,0,1,2, etc., and these parts can be divided into smaller parts with any desired precision. The resulting coordinates may or may not be orthogonal.

Assuming the surface is immersed in a three-dimensional Euclidean space with coordinates (X, Y, Z), we can express its coordinates using the following equations:

$$X = X(x,y), \quad Y = Y(x,y), \quad Z = Z(x,y) \tag{9-1}$$

Where we assume they are differentiable as many times as desired. For example, for a sphere with radius (a) whose center is located at the coordinate origin, we have

$$X = a\sin x \cos y, \quad Y = a\sin x \sin y, \quad X = a\cos x$$

The distance between adjacent points in Euclidean space is given by:

$$d\sigma^2 = dX^2 + dY^2 + dZ^2 \tag{9-2}$$

This allows us to calculate the distance on the surface using

$$d\sigma^2 = \left(X_1 dx + X_2 dy\right)^2 + \left(Y_1 dx + Y_2 dy\right)^2 + \left(Z_1 dx + Z_2 dy\right)^2 \tag{9-3}$$

Here, the subscripts 1 and 2 show the partial derivative with respect to x and y, respectively. It is clear that (9-3) can be written as following:

$$d\sigma^2 = Edx^2 + 2Fdxdy + Gdy^2 \tag{9-4}$$

where G, F and E are functions of x and y. (This method in the case of sphere, reaches the following relation: $d\sigma^2 = a^2 dx^2 + a^2 \sin^2 x dy^2$ which can be obtained directly with the help of elementary geometry). Whenever the square of the distance differential $d\sigma^2$ is given by a homogeneous quadratic differential equation in terms of surface coordinates, such as (9-4), we say that $d\sigma^2$ is a Riemannian metric and the corresponding surface is a Riemannian surface. Of course, all metrics do not necessarily have to be like this. The distinction of Riemannian metric from other metrics is that, Riemannian geometry is locally Euclidean. At each assumed point p_0, the quantities G, F, E in (9-4) are just numbers, which we call them, for example E_0, F_0, G_0, so we write $d\sigma^2$ in the point p_0 as a sum of several squares

$$d\sigma^2 = \left(E_0^{1/2}dx + \frac{F_0}{E_0^{1/2}}dy \right)^2 + \left(G_0 - \frac{F_0^2}{E_0} \right)dy^2 = d\tilde{x}^2 + d\tilde{y}^2 \qquad (9\text{-}5)$$

Where $\quad \tilde{x} = E_0^{1/2}x + \dfrac{F_0}{E_0^{1/2}}y, \quad \tilde{y} = \left(G_0 - \dfrac{F_0^2}{E_0} \right)^{1/2}y$

 So there is a coordinate transformation (actually infinity of transformations) that makes the metric "Euclidean" at any preselected point (that is, the set of poly differential squares). On the contrary, if there are coordinates such as $\tilde{x}, \tilde{y}$ that the metric becomes Euclidean at a point p_0 according to it, this metric at the point p_0 in terms of general coordinates must be Riemannian, because a transformation such as $\tilde{x} = \tilde{x}(x,y), \tilde{y} = \tilde{y}(x,y)$ from special coordinates must exist, and therefore

$$d\tilde{x}^2 + d\tilde{y}^2 = \left(\tilde{x}_1 dx + \tilde{x}_2 dy \right)^2 + \left(\tilde{y}_1 dx + \tilde{y}_2 dy \right)^2 \qquad (9\text{-}6)$$

which is Riemannian in terms of x and y. Now we see that, to predict the form of the formula (9-4), it was enough to pay attention that the surface is locally Euclidean, that is, for each assumed point, the coordinate lines can be drawn in such a way that in that point, $d\sigma^2 = dx^2 + dy^2$.

These ideas can be directly extended from the surface to higher dimensional spaces.

Now it is the turn of a generalization which, although very brief in form, but is important. This generalization consists in allowing metrics that are not positive definite meaning they can have mixed signs. In many cases, the condition that all terms be positive, such as G>0, F>0, E>0, is not necessary. It is clear that on a real surface for every dx and $dy \neq 0$, we have $d\sigma^2 > 0$, but this condition is not essential for most part of the theory. Every indeterminate Riemannian metric locally corresponds to a pseudo-Euclidean metric (the metric only contains the square of the differentials, but the sign of some of them is negative), for example, we have corresponding to the signature (+,+,-): $dx^2 + dy^2 - dz^2$ it can be shown that the signature of a metric is invariant, i.e. Assuming that the coordinates are real, no matter how the metric is reduced to the sum of squared differentials, the distribution of positive and negative signs will always be the same. An example for the Riemannian space with the signature (+,-,-,-), is the space-time of special relativity (hereafter we call it M_4 Minkowski space), with metric:

$$ds^2 = c^2 dt^2 - dx^2 - dy^2 - dz^2$$

Of course, this is a special example because this space is not only Riemannian, but everywhere pseudo-Euclidean. The fact that ds here is not a simple ruler distance, does not affect the mathematical expression of the matter. In this case, as in the case of positive definite spaces, the differential equation of geodesics is obtained from the condition of length $\sigma = \int |ds| = 0$, but in indefinite spaces, geodesics are no longer curves of minimum length, but generally have neighbors of greater or less length. Geodesics of maximum length, that is, geodesics for which $ds^2 > 0$ and $[<0]$, exist only when the signature has only one positive (negative) sign. The sign of ds^2 along the geodesics in each Euclidean space correspond to linear equations in terms of Euclidean coordinates. The differential equation of geodesics can also be interpreted as following:

Geodesics are "locally straight", that is, their curvature in the Euclidean space of the local tangent is zero. From here, the following experimental method for finding geodesics of a surface is obtained: we cut a thin strip of paper and draw a straight line along its middle. Then we stick the tape piece by piece without removing it carefully on the surface. The drawn line will be a geodesic on the surface. By knowing the metric, all distance relations in space are known and from there, what is needed for the intrinsic knowledge of space is obtained. For example, the differential equation of geodesics includes only coordinates and metric coefficients (that is, the generalization of coefficients G, F, E in (9-4)).

Certainly, in both spaces that are essentially equivalent, it is possible to define coordinates that metrics to be the same according to it (it is enough to match two spaces and map the coordinates of one on the other). On the contrary, if the metrics can be made the same by choosing appropriate coordinates, the spaces are intrinsically equivalent. (By pairing corresponding points, one can be mapped onto the other) Hence, the shorter word "isometric" is used instead of "intrinsically equivalent". This result brings us to an important issue. Two spaces that appear to have completely different metrics can be isometric. For example, from the following four metrics

$$dx^2 + x^2 dy^2, \ \left(4x^2 + y^2\right)dx^2 + \left(2xy - 4x\right)dxdy + \left(1 + x^2\right)dy^2$$

$$y^2 dx^2 + x^2 dy^2, \ \ ydx^2 + xdy^2 \tag{9-7}$$

The first three metrics are representative of a normal page, but the last one is not. The former is actually polar regular metric $dr^2 + r^2 d\theta$ with unconventional notation, but still recognizable. The second one is obtained from the normal Euclidean coordinates $\tilde{y}, \tilde{x}$ by the following unsigned transformation $\tilde{x} = x^2 - y$, $\tilde{y} = xy$

The third one is obtained from the transformation of Euclidean coordinates.

Therefore interestingly, two spaces which appear to have entirely different metrics can still be isometric. The metrics provided in (9-7) illustrate this concept, where they may seem distinct but can be transformed to be equivalent in the right coordinates.

10- Einstein field equations

The Einstein field equations which are the cornerstone of general relativity, establish the connection between the curvature of space-time and the distribution of matter and energy. The equation is represented as:

$$R_{\mu\upsilon} - \frac{1}{2}g_{\mu\upsilon}R = -kT_{\mu\upsilon}$$

(10-1)

Where, $R_{\mu\upsilon}$ is the Ricci curvature tensor, R is the curvature scalar, $g_{\mu\upsilon}$ is the metric tensor and $T_{\mu\upsilon}$, is the energy-momentum tensor characterizing the distribution of matter and energy within space-time.

When the energy-momentum tensor assumes a value of zero, the Einstein field equation reduces to:

$$R_{\mu\upsilon} = 0$$

(10-2)

This form, known as the vacuum field equation, describes the curvature of space-time in the absence of matter and energy. The Einstein vacuum field equation has fundamental concept in general relativity, emphasizing that the curvature of space-time is determined by the presence and distribution of matter and energy within it. In other words, matter and energy create the curvature of space-time, which then influences the motion of matter and energy itself. This interaction is encapsulated in the general form of the Einstein field equations, as described in the following form:

$$G_{\mu\upsilon} = -kT_{\mu\upsilon}$$

(10-3)

Where $G_{\mu\upsilon}$ is the Einstein tensor, which is, a combination of the Ricci curvature tensor and the metric tensor

$$G_{\mu\upsilon} = R_{\mu\upsilon} - \frac{1}{2} g_{\mu\upsilon} R$$

(10-4)

Equation (10-3) signifies the interplay between the curvature of space-time and the energy-momentum tensor, thereby explaining gravitational effects.

11- Differential geometry elements

In Differential geometry we deal with manifolds. Basically manifolds are spaces that are locally Euclidean.

We know that an n-dimensional Euclidean space, R_n, is the set of all n-tuples with closed and open sets, $(x^1,....,x^n)$ $(-\infty\langle x^i \langle +\infty)$.

Locally, a manifold, M, is the same as Euclidcan space, in the sense that, M is covered (i.e., a union of) neighborhoods, u_α, and related to each υ_α there's a one to one map, φ_α, which it images each point $p \in u_\alpha$, to a point in an open neighborhood of R_n with the coordinates $(x^1,....,x^n)$.

If we multiply two manifolds M and N, the Cartesian product $M \times N$ of them will be, the ordered pair of points, (p, q), where $p \in M$ and $q \in N$; meanwhile, if u_α, and v_α, are neighborhoods in M and in N, φ_α and ψ_α are the associated maps, and $\varphi_\alpha(p) = (x^1,....,x^n)$ and

$\psi_\beta(q)=(y^1,....,y^m)$ (m is not equal to n necessarily), then for defining

$M \times N$ as a manifold of (m+ n) dimensions, it suffices to use the following map to complete the definition,

$$(\varphi_\alpha \times \psi_\beta)(p,q)=(x^1,.....,x^n,y^1,.....,y^m)$$

Now, we consider a function f on M, which is, defined by a map $f : M \rightarrow R^1$. We must assume that the combined map $f \circ \varphi_\alpha^{-1}$ is a smooth function of the coordinates $(x^1,.....,x^n)$, which it images a point $(x^1,.....,x^n)$ in R^n on to the reals, R^1. Here a smooth curve λ on M, can be defined by the map $\lambda :$ an interval I (a <t < b) in $R^1 \rightarrow \lambda(t)=p \in M$

Such that

$$(\varphi_\alpha \circ \lambda)(t)=\left[x^1(t),.....,x^n(t)\right];\qquad (11\text{-}1)$$

and it is required that $x^i(t)(i=1,...,n)$ are smooth functions of t.

We must pay attention that a function f, which is defined on the manifold, helps us to define the function, $f \circ \lambda$ on the curve λ. We can consider the function $f(\lambda(t))=f(x^1(t),...,x^n(t))$ using the map $\varphi_\alpha \circ \lambda$, where $(x^1(t),...,x^n(t))$ are the coordinates of $p=\lambda(t)$ by the map φ_α.

(a) Tangent vectors

On a curve λ on M, using the definition of $f(x^1(t),...,x^n(t))$, that is given above, consider

$$\left(\frac{\partial f}{\partial t}\right)_{\lambda(t)}\Bigg|_{t=t_0} = \lim_{\varepsilon \to 0} \frac{1}{\varepsilon}\{f(\lambda(t_0+\varepsilon))-f(\lambda(t_0))\}$$

$$= \sum_{j=1}^{n}\frac{dx^j(t)}{dt}\Bigg|_{t_0}\left(\frac{\partial f}{\partial x^j}\right)_{\lambda(t_0)} = \left(\frac{dx^j}{dt}\frac{\partial f}{\partial x^j}\right)_{t_0} \qquad (11\text{-}2)$$

where the summation is done on the repeated indices.

Now it is obvious that, using different curves which they are passing through a point p, a linear vector space at p can be defined, which is consisting of linear combinations of the coordinate derivatives $\partial/\partial x^j$ of the forms

$$X = X^j\frac{\partial}{\partial x^j} \qquad (11\text{-}3)$$

where x^j are any arbitrary set of n numbers. For t (which is in a small interval $-\varepsilon < t < +\varepsilon$), by considering the curves λ which is defined by

$$x^j(t) = x^j(p) + x^j t \qquad (j=1, ..., n) \qquad (11\text{-}4)$$

the above tangent vectors in (11-3) arise.

Since the necessity for a linear vector space,

$$(\alpha X + \beta Y)f = \alpha(Xf) + \beta(Yf) \qquad (11\text{-}5)$$

is satisfied for all X, Y, α, β and f, the tangent vectors at a given point p, form a linear vector space on R^1 spanned by the derivatives of coordinate.

If we define X by

$$Xf = X^j \frac{\partial f}{\partial x^j} = X^j f_{,j} \qquad (11\text{-}6)$$

then for every smooth function f, when it operates on products of functions, expression (11-6) will satisfy the Leibnitz rule, so we have

$$X(fg)\big|_{\lambda(t)} = (fXg + gXf)\big|_{\lambda(t)} \qquad (11\text{-}7)$$

In fact, we may consider the tangent vectors as directional derivatives (in eq. (11-6) it has been introduced the notation of denoting derivatives with respect to x^j by the index j, which it follows by a comma, both as subscripts.).

 The tangent vector's space (which also they are called contravariant vectors) to a n dimensional manifold M at p, signify by $T_p(M)$ or simply T_p, is an n-dimensional vector-space. May be we want to visualize this space as the set of all directions at p, which in this case, this space is called, the tangent space at p.

We can choose n linearly independent vectors e_a (a = 1, . . . , n), instead of a basis determined by local coordinates. Then it must be exist linear relations of the form:

$$e_a = \phi_a^k \frac{\partial}{\partial x^k}$$

$$(11\text{-}8)$$

where if we calculate the determinant of the matrix that is formed by ϕ_a^k , it will be seen that this determinant must be nonzero. Then the inverse relation will be:

$$\frac{\partial}{\partial x^j} = \phi_j^b e_b$$

$$(11\text{-}9)$$

Thus the inverse of the matrix $\left[\phi_a^k \right]$ is $\left[\phi_j^b \right]$:

$$\phi_a^k \phi_j^a = \delta_j^k \quad \text{and} \quad \phi_a^k \phi_k^b = \delta_a^b .$$

$$(11\text{-}10)$$

Any tangent vectors at p , for any basis (e_j), can be expressed in the form

$$X = X^j e_j .$$

$$(11\text{-}11)$$

Where relative to the (e_j), the X^{j}'s are the components of X.

(b) One-forms or covariant vectors

If we suppose that, ω is a one-form at p and T_p is the tangent space at p, then ω will be a linear mapping of T_p on to the reals:

$$\omega: \quad T_p \rightarrow R^1$$

$$(11\text{-}12)$$

In the other hand, a number $\omega(X)$, which is associated uniquely to the one-form ω for any tangent vector X at p, can be written as:

$$\omega(X) = \langle \omega, X \rangle. \tag{11-13}$$

We can express the linearity of the map by the relation:

$$\langle \omega, \alpha X + \beta Y \rangle = \alpha \langle \omega, X \rangle + \beta \langle \omega, Y \rangle, \tag{11-14}$$

where α and β are any arbitrary real numbers and X , Y are two tangent vectors.

If ω and π are two one-forms, for any $X \in T_p$ and any real number α, we can define the multiplication of forms by real numbers and sums of forms as:

$$(\alpha\omega)(X) = \alpha \langle \omega, X \rangle \quad \text{and} \quad (\omega + \pi)(X) = \langle \omega, X \rangle + \langle \pi, X \rangle, \tag{11-15}$$

Using these rules, one-forms span a vector space T_p^*. We call T_p^*, the dual of the tangent space or cotangent space at p. Thus because of this, we call cotangent vectors (or covariant vectors) to the one-forms.

Now we should consider that the one-forms $(e^i)(i = 1, \ldots, n)$ which map any tangent vector $X = X^j e_j$ to its components, provide a basis for T_p^*, associated with a basis (e_j) for T_p , thus,

$$e^i(X) = \langle e^i, X^j e_j \rangle = X^i \qquad (i = 1,..., n). \tag{11-16}$$

It follows from equation (11-16) that,

$$e^i(e_j) = \langle e^i, e_j \rangle = \delta_j^i. \tag{11-17}$$

By observing:

44

$$\langle \omega, X \rangle = \langle \omega, X^i e_i \rangle = X^i \langle \omega, e_i \rangle . \tag{11-18}$$

we obtain that any one-form ω, can be expressed as a linear combination of the e^i's .

Let the ω_i be the numbers that, ω maps the basis vectors (e_i) of the tangent space T_p at p, to them

$$\omega_i = \langle \omega, e_i \rangle = \omega(e_i) \tag{11-19}$$

So we can write:

$$\langle \omega, X \rangle = \omega_i X^i = \omega_i \langle e^i, X^j e_j \rangle = \langle \omega_i e^i, X \rangle . \tag{11-20}$$

Since equation (11-20) is valid for any $X \in T_p$, we have

$$\omega = \omega_i e^i ; \tag{11-21}$$

And this is the expression that is required for ω as a linear combination of the e^i. From its definition, it is clear that the vectors e^i are linearly independent. We can say, the bases (e_i) and (e^i) provide dual bases for the cotangent and tangent spaces at p.

If we choose the following different bases instead of the dual bases (e_i) and (e^j),

$$e_{i'} = \phi_{i'}^j e_j \quad \text{and} \quad e^{j'} = \phi_j^{j'} e^j \tag{11-22}$$

where they are obtained by non-singular linear transformations demonstrated by $\phi_{i'}^j$ and $\phi_j^{j'}$, then for the purpose that the duality of the bases $(e_{i'})$ and $(e^{j'})$ to be continued, we require the following condition:

$$\delta_{i'}^{j'} = \left\langle e^{j'}, e_{i'} \right\rangle = \phi_j^{j'} \phi_{i'}^k \left\langle e^j, e_k \right\rangle$$
$$= \phi_j^{j'} \phi_{i'}^k \delta_k^j = \phi_j^{j'} \phi_{i'}^j$$

(11-23)

strictly speaking, the matrices $[\phi_{i'}^i]$ and $[\phi_j^{j'}]$ are the inverses of each other.

If one set of coordinates (x^i) is changed to another set $(x^{i'})$, then we can write the following expressions for $\phi_{i'}^i$ and $\phi_j^{j'}$

$$\phi_{i'}^i = \left(\frac{\partial x^i}{\partial x^{i'}} \right)_p \qquad \text{and} \qquad \phi_j^{j'} = \left(\frac{\partial x^{j'}}{\partial x^j} \right)_p \qquad (11\text{-}24)$$

Now we define a one-form df, related to any function f on the manifold by requiring that:

$$df(X) = \left\langle df, X \right\rangle = Xf, \qquad (11\text{-}25)$$

where $X \in T_p$. In a local coordinate basis,

$$X = X^i \frac{\partial}{\partial x^i} \qquad (11\text{-}26)$$

and,

$$\left\langle df, X \right\rangle = X^i \frac{\partial f}{\partial x^i} = X^i f_{,i}; \qquad (11\text{-}27)$$

and,

$$\left\langle dx^{\,j}, \frac{\partial}{\partial x^{\,i}} \right\rangle = \delta_i^{\,j} \; ; \qquad\qquad (11\text{-}28)$$

Thus, a local coordinate basis for the cotangent vectors provided by the one-forms $(dx^{\,j})$, is dual to the local coordinate basis which is provided by the tangent vectors $(\partial_j = \frac{\partial}{\partial x_{\,j}})$ for the tangent space. Sometimes we refer to the bases (∂_j) and $(dx^{\,j})$ as the canonical bases for the cotangent and the tangent spaces.

We must pay attention that if,

$$df = \alpha_j dx^{\,j} , \qquad\qquad (11\text{-}29)$$

then it is concluded from

$$\begin{aligned} X^{\,i} f_{,i} &= \left\langle df, X \right\rangle = \left\langle \alpha_j dx^{\,j}, X^{\,i} \partial_i \right\rangle \\ &= \alpha_j X^{\,i} \left\langle dx^{\,j}, \partial_i \right\rangle = \alpha_i X^{\,i} \end{aligned} \qquad (11\text{-}30)$$

that

$$\alpha_i = f_{,i} \qquad \text{and} \qquad df = f_{,i} dx^{\,i} \qquad (11\text{-}31)$$

and this equation is consistent with the ordinary meaning of df.

(c) Tensor products

The following expression

$$\Pi_r^s = \underbrace{T_p^* \times T_p^* \times .... \times T_p^*}_{r \; factors} \times \underbrace{T_p \times T_p \times .... \times T_p}_{s \; factors} \qquad (11\text{-}32)$$

illustrate the Cartesian product or s tangent spaces and r cotangent spaces at point p of a manifold, i.e., ordered sets space of s tangent vectors and r one-forms:

$$(\omega^1, \ldots, \omega^r, X_1, \ldots, X_s)$$

One can consider a multilinear mapping, T, of the Π_r^s manifold, to the reals:

$$T: \qquad \Pi_r^s \rightarrow R^1 \qquad (11\text{-}33)$$

The mapping provides an association of any ordered set of s tangent vectors and r one-forms to a real number:

$$T(\omega^1, \ldots, \omega^r, X_1, \ldots, X_s) = \text{ a real number.} \qquad (11\text{-}34)$$

If we want the map to be multilinear, it is required that,

$$T(\omega^1, \ldots, \omega^r, \alpha X + \beta Y, X_2, \ldots, X_s) =$$

$$= \alpha \, T(\omega^1, \ldots, \omega^r, X, X_2, \ldots, X_s) + \beta \, T(\omega^1, \ldots, \omega^r, Y, X_2, \ldots,$$

$$X_s) \quad (11\text{-}35)$$

for all X, $Y \in T_p$ and $\alpha, \beta \in R^1$; and for analogous substitutions of all the other forms and vectors. Now we define the multilinear mapping. A multilinear mapping is a tensor of type (r, s).

For $\omega^i \in T_p^*$, and $\alpha, \beta \in R^1$ and $X_j \in T_p$ (i = 1, . . . ,r; j = 1, . . . ,s), we define Linear combinations of tensors of type (r, s) by the rule

$$(\alpha T + \beta S)(\omega^1, \ldots, \omega^r, X_1, \ldots, X_s) =$$

$$= \alpha\, T(\omega^1, \ldots, \omega^r, X_1, \ldots, X_s) + \beta\, S(\omega^1, \ldots, \omega^r, X_1, \ldots, X_s)$$

$$(11\text{-}36)$$

We call to the space of such tensors, the space of tensor products, and denote them by

$$T_s^r(p) = \underbrace{T_p \otimes \ldots \otimes T_p}_{r\ factors} \otimes \underbrace{T_p^* \otimes \ldots \otimes T_p^*}_{s\ factors} \qquad (11\text{-}37)$$

Now we must verify that we need n^{r+s} special mappings to provide a basis for tensor products of type (r, s)

$$e_{i_1 \ldots i_r}^{j_1 \ldots j_s}(\omega^1, \ldots, \omega^r, X_1, \ldots, X_s)$$

$$= e_{i_1 \ldots i_r}^{j_1 \ldots j_s}(\omega_{k_1}^1 e^{k_1}, \ldots, \omega_{k_r}^r e^{k_r}, X_1^{l_1} e_{l_1}, \ldots, X_s^{l_s} e_{l_s}) \qquad (11\text{-}38)$$

These mappings are tensors of type (r,s) and they are linear in every argument. For defining these mappings, we use the following equivalent way

$$e_{i_1 \ldots i_r}^{j_1 \ldots j_s} (e^{k_1}, \ldots, e^{k_r}, e_{l_1}, \ldots, e_{l_s}) = \delta_{i_1}^{k_1} \delta_{i_2}^{k_2} \ \ldots \ \delta_{i_r}^{k_r} \delta_{l_1}^{j_1} \ \ldots \ \delta_{l_s}^{j_s} \quad (11\text{-}39)$$

We can express all type (r, s) tensors as a linear combination of the mappings (11-39) that follows from noting

$$T(\omega^1, \ldots, \omega^r, X_1, \ldots, X_s) = (\omega_{i_1}^1 e^{i_1}, \ldots, \omega_{i_r}^r e^{i_r}, X_1^{j_1} e_{j_1}, \ldots,$$

$$X_s^{j_s} e_{j_s})$$

$$= \omega_{i_1}^1, \ldots, \omega_{i_r}^r X_1^{j_1} \ldots X_s^{j_s} T(e^{i_1}, \ldots, e^{i_r}, e_{j_1}, \ldots, e_{j_s}) \qquad (11\text{-}40)$$

and

$$T(e^{i_1}, \ldots, e^{i_r}, e_{j_1}, \ldots, e_{j_s}) = T_{j_1 \ldots j_s}^{i_1 \ldots i_r} \qquad (11\text{-}41)$$

Then we can write:

$$T = T_{j_1 \ldots j_s}^{i_1 \ldots i_r} e_{i_1 \ldots i_r}^{j_1 \ldots j_s} \qquad (11\text{-}42)$$

and this is the expression that we required for T as a linear combination of the mappings (11-39).

It is clear that the mappings (11-39) are independent linearly. So, they provide a basis for type (r, s) tensors. These basis elements $(e_{i_1 \ldots i_r}^{j_1 \ldots j_s})$ number is n^{r+s}.

Relative to the chosen basis, The coefficients, $T_{j_1 \ldots j_s}^{i_1 \ldots i_r}$, in equation (11-42), are the components of T.

We can write:

$$e^{j_1\cdots j_s}_{i_1\cdots i_r} = e_{i_1} \otimes \ldots \otimes e_{i_r} \otimes e^{j_1} \otimes \ldots \otimes e^{j_s} \qquad (11\text{-}43)$$

as indicating the dual bases tensor product (e_i) and (e^j) of T_p and T_p^*.

The following tensor product of s one-forms and r tangent vectors

$$Y_1 \otimes \ldots Y_r \otimes \Omega^1 \otimes \ldots \otimes \Omega^s \qquad (11\text{-}44)$$

is that T_s^r element which maps $(\omega^1, \ldots, \omega^r, X_1, \ldots, X_s)$ to the number

$$\langle \omega^1, Y_1 \rangle \ldots \langle \omega^r, Y_r \rangle \langle \Omega^1, X_1 \rangle \ldots \langle \Omega^s, X_s \rangle \qquad (11\text{-}45)$$

Now we can wrtite:

$$(e_{i_1} \otimes \ldots \otimes e_{i_r} \otimes e^{j_1} \otimes \ldots \otimes e^{j_s})(\omega^1, \ldots, \omega^r, X_1, \ldots, X_s)$$

$$= \langle \omega^1, e_{i_1} \rangle \ldots \langle \omega^r, e_{i_r} \rangle \langle e^{j_1}, X_1 \rangle \ldots \langle e^{j_s}, X_s \rangle$$

$$= \omega^1_{i_1} \ldots \omega^r_{i_r} X_1^{j_1} \ldots X_s^{j_s} = e^{j_1\cdots j_s}_{i_1\cdots i_r}(\omega^1, \ldots, \omega^r, X_1, \ldots, X_s) \qquad (11\text{-}46)$$

If we choose different dual bases $(e_{i'})$ and $(e^{j'})$, in place of the dual bases (e_i) and (e^j), then from equations (11-22), we conclude that relative to the new basis,

$$e_{i_1'} \otimes \ldots \otimes e_{i_r'} \otimes e^{j_1'} \otimes \ldots \otimes e^{j_s'}$$

the components of T are given by:

$$T^{i_1'\cdots i_r'}_{j_1'\cdots j_s'} = \phi^{i_1'}_{i_1} \ldots \phi^{i_r'}_{i_r} \phi^{j_1}_{j_1'} \ldots \phi^{j_s}_{j_s'} T^{i_1\cdots i_r}_{j_1\cdots j_s} \qquad (11\text{-}47)$$

Suppose that, we have a type (r,s) tensor with the components $T_{j_1...j_s}^{i_1...i_r}$, if we calculate the contraction of this tensor with respect to a chosen covariant index j_q and contravariant index i_p , we will find the following tensor of type (r - 1, s - 1):

$$T_{j_1...j_{q-1}kj_{q+1}...j_s}^{i_1...i_{p-1}ki_{p+1}...i_r} e_{i_1} \otimes ... \otimes e_{i_{p-1}} \otimes e_{i_{p+1}} \otimes ... \otimes e_{i_r}$$
$$\otimes e^{j_1} \otimes ... \otimes e^{j_{q-1}} \otimes e^{j_{q+1}} \otimes ... \otimes e^{j_s} \tag{11-48}$$

where, the summation on all values of k (of the i_p –contravariant index and j_q -covariant index) must be effected. Using equations (11-23) and (11-47), we can easily verify that the process of contraction is independent of the chosen basis.

We say symmetric or antisymmetric to a tensor of type (0,2) if, for all X and Y in T_p , we have:

$$T(X, Y) = T(Y, X) \qquad \text{or} \qquad T(X, Y) = -T(Y, X) \tag{11-49}$$

In an arbitrary basis, in components terms, we see that the antisymmetry or symmetry indicates that,

$$T_{ij} = T_{ji} \qquad \text{or} \qquad T_{ij} = -T_{ji} \tag{11-50}$$

Generally, we say symmetric or antisymmetric to a type (r,s) tensor, in its covariant indices i and j, if,

$$\text{T}(\omega^1, ..., \omega^i, \ \omega^j, ..., \omega^r, \ X_1, \ ..., \ X_s) =$$

$$\pm T(\omega^1, ..., \omega^j, \ \omega^i, ..., \omega^r, \ X_1, \ ..., \ X_s) \tag{11-51}$$

for all X's and ω's . We can define similarly the Symmetry or antisymmetry with respect to contravariant indices.

12. Antisymmetric tensors and forms

The class of totally antisymmetric tensors is an important class of tensors of type (0, s). This class contains covariant tensors with antisymmetric property in every pair of their arguments, i.e.

$$T(X_1,...,X_i,...,X_j,...,X_s) = -T(X_1,...,X_j,...,X_i,...,X_s),\qquad(12\text{-}1)$$

for all pairs of indices i and j and for all X's. By applying the alternating operator A to a general tensor T of type (0,s), this kind of tensor can be formed. Applying operator A to T gives a linear combination defined by:

$$AT(X_1,...,X_s) = \frac{1}{s!}\sum_{j_1,...,j_s} \text{sgn}(j_1,...,j_s)T(X_{j_1},...,X_{j_s})\qquad(12\text{-}2)$$

where in this summation, $(j_1,...,j_s)$ is an even or an odd permutation of $(1,...,s)$ integer numbers, and, $\text{sgn}(\upsilon_1,...,\upsilon_s)=\pm1$, and equation (12-2) is to be valid for every $(X_1,...,X_s)$.

When T is totally antisymmetric, applying the A operator to it simply reproduces T. However, when $s\rangle n$ (the dimension of the vector space), applying the A operator to T reduces T to zero; simply put, there is no totally antisymmetric tensor of type (0, s) for $s\rangle n$.

We call s-forms to the antisymmetric tensors of type (0,s). S-forms span a vector space which we represent this space by $\Lambda^s T_p^*$.

By applying the A operator to the basis elements of the following tensor product,

$$A(e^{j_1} \otimes ... \otimes e^{j_s})$$

(12-3)

we can obtain a basis for $\Lambda^s T_p^*$.

The resulting basis elements can be written as the exterior or the wedge product of the e^j's as the following:

$$e^{j_1} \wedge e^{j_2} \wedge .... \wedge e^{j_s} \quad (j_1 \rangle j_2 \rangle .... \rangle j_s)$$

(12-4)

By extending the summation only over strictly descending sequences, a general s-form can be written as:

$$\Omega = \Omega_{j_1...j_s} e^{j_1} \wedge e^{j_2} \wedge .. e^{j_s} \quad (j_1 \rangle j_2 \rangle .... \rangle j_s)$$

(12-5)

Considering that, interchanging a pair of indices is equal to interchanging the corresponding elements in the wedge product, it can be deduced that interchanging the elements in a wedge product must be accompanied by a change of sign:

$$e^j \wedge e^k = -e^k \wedge e^j$$

(12-6)

The expression for an s-form in a local coordinate basis is:

$$\Omega = \Omega_{j_1...j_s} dx^{j_1} \wedge \ ... \ \wedge dx^{j_s}$$

(12-7)

To obtain a (p+q) form, the wedge product of any p-form Ω^1 and a q-

54

form Ω^2 can be formed by the rule

$$\Omega^1 \wedge \Omega^2 = A(\Omega^1 \otimes \Omega^2) \qquad (12\text{-}8)$$

which must accordingly vanish identically if $(p+q)\rangle n$.

By definition,

$$\Omega^1 \wedge \Omega^2 = (\Omega^1_{j_1 \dots j_p} e^{j_1} \wedge \; \dots \; \wedge e^{j_p}) \wedge (\Omega^2_{k_1 \dots k_p} e^{k_1} \wedge \; \dots \; \wedge e^{k_p}) \quad (12\text{-}9)$$

where $(j_1, \dots, j_p)$ and $(k_1, \dots, k_q)$ are strictly descending sequences. Since each of the q basis elements $e^{k_1}, \dots, e^{k_q}$ must go through p interchanges before $\Omega^1 \wedge \Omega^2$ can be brought to the form required of $\Omega^2 \wedge \Omega^1$, consequently:

$$\Omega^1 \wedge \Omega^2 = (-1)^{pq} (\Omega^2_{k_1 \dots k_p} e^{k_1} \wedge \; \dots \; \wedge e^{k_q}) \wedge (\Omega^1_{j_1 \dots j_p} e^{j_1} \wedge \; \dots \; \wedge e^{j_p})$$
$$= (-1)^{pq} \Omega^2 \wedge \Omega^1$$

$$(12\text{-}10)$$

13. Exterior differentiation

When exterior differentiation is applied to forms, it converts p-forms to (p+ l)-forms. Exterior differentiation is effected by an operator d with the following rules:

(a) If we apply the operator d to functions (or zero-forms) f, it yields a one-form d/ defined by

$$df(X) = \langle df, X \rangle = Xf \tag{13-1}$$

for every $X \in T_0^1$. Mainly in a local coordinate basis we have:

$$df = \frac{\partial f}{\partial x^j} dx^j \tag{13-2}$$

(b) For two p-forms, A_1 and A_2 we have:

$$d(\alpha A_1 + \beta A_2) = \alpha dA_1 + \beta dA_2 \qquad (\alpha, \beta \in ¡^1) \tag{13-3}$$

(c) If we suppose that, A is a p-form and B is a q-form, we have,

$$d(A \wedge B) = dA \wedge B + (-1)^p A \wedge dB \tag{13-4}$$

(d) For every p-form A, Poincare's lemma requires that, $d(dA) = 0$

For considering that the operator d in the foregoing rules is well-defined, first we consider the exterior derivative of a p-form A,

$$dA = d(A_{j_1 \cdots j_p} dx^{j_1} \wedge \ldots \wedge dx^{j_p})$$

Using (a), (b), and (d) rules we have,

$$dA = dA_{j_1 \ldots j_p} \wedge dx^{j_1} \wedge \ldots \wedge dx^{j_p})$$

$$= \frac{\partial A_{j_1 \ldots j_p}}{\partial x^{k}} dx^{k} \wedge dx^{j_1} \wedge \ldots \wedge dx^{j_p} \qquad (13\text{-}5)$$

Here the important point is that, we must verify that dA, which is given in equation (13-5), is independent of the choice of the local coordinate system. If we had chosen a different set of local coordinates ($x^{j'}$), in place of coordinates (x^{j}), then using equations (11-24) and (11-47),

$$A_{j_1' \ldots j_p'} = A_{j_1 \ldots j_p} \frac{\partial x^{j_1}}{\partial x^{j_1'}} \ldots \frac{\partial x^{j_p}}{\partial x^{j_p'}} \qquad (13\text{-}6)$$

and we infer that

$$d(A_{j_1' \ldots j_p'} dx^{j_1'} \wedge \ldots \wedge dx^{j_p'})$$

$$= d(A_{j_1 \ldots j_p} \frac{\partial x^{j_1}}{\partial x^{j_1'}} \ldots \frac{\partial x^{j_p}}{\partial x^{j_p'}} dx^{j_1'} \wedge \ldots \wedge dx^{j_p'}) = .$$

$$= \frac{\partial x^{j_1}}{\partial x^{j_1'}} \ldots \frac{\partial x^{j_p}}{\partial x^{j_p'}} dA_{j_1 \ldots j_p} \wedge dx^{j_1'} \wedge \ldots \wedge dx^{j_p'} +$$

$$+ \frac{\partial^2 x^{j_1}}{\partial x^{k'} \partial x^{j_1'}} \frac{\partial x^{j_2}}{\partial x^{j_2'}} \ldots \frac{\partial x^{j_p}}{\partial x^{j_p'}} A_{j_1 \ldots j_p} dx^{k'} \wedge dx^{j_1'} \wedge \ldots \wedge dx^{j_p'} + \ldots$$

$$\ldots + \frac{\partial x^{j_1}}{\partial x^{j_1'}} \ldots \frac{\partial x^{j_{p-1}}}{\partial x^{j_{p-1}'}} \frac{\partial^2 x^{j_p}}{\partial x^{k'} \partial x^{j_p'}} dA_{j_1 \ldots j_p} dx^{k'} \wedge dx^{j_1'} \wedge \ldots \wedge dx^{j_p'} \qquad (13\text{-}7)$$

If we consider the right-hand side of equation (13-7), we will see that all terms on the right-hand which involve the second derivatives of x^{j_i} (i = 1, . . . ,p) will be zero, because of the terms symmetry in k' and j_i' and the antisymmetry of the basis elements, in the same indices in the wedge product. The first term is the only surviving term which is the same as

$$dA_{j_1...j_p} \wedge dx^{j_1} \wedge \ ... \ \wedge dx^{j_p} = d(A_{j_1...j_p} dx^{j_1} \wedge \ ... \ \wedge dx^{j_p})$$

As a result,

$$dA_{j'_1...j'_p} \wedge dx^{j'_1} \wedge \ ... \ \wedge dx^{j'_p} = d(A_{j_1...j_p} dx^{j_1} \wedge \ ... \ \wedge dx^{j_p}) \qquad (13\text{-}8)$$

that we set out to verify this.

We next consider that, for dA, the rule (c) is consistent with the phrase (13-5).

Using rules (a), (b), and (d)

$$d(A \wedge B) = d(A_{j_1...j_p} dx^{j_1} \wedge \ ... \ \wedge dx^{j_p} \wedge B_{k_1...k_q} dx^{k_1} \wedge \ ... \ \wedge dx^{k_q}$$

$$= \frac{\partial A_{j_1...j_p}}{\partial x^i} dx^i \wedge dx^{j_1} \wedge \ ... \ \wedge dx^{j_p} \wedge B_{k_1...k_q} dx^{k_1} \wedge \ ... \ \wedge dx^{k_q}) +$$

$$+ A_{j_1...j_p} \frac{\partial B_{k_1...k_q}}{\partial x^i} dx^i \wedge dx^{j_1} \wedge \ ... \ \wedge dx^{j_p} \wedge dx^{k_1} \wedge \ ... \ \wedge dx^{k_q} \qquad (1$$

$$= dA \wedge B + (-1)^p A_{j_1...j_p} dx^{j_1} \wedge \ ... \ \wedge dx^{j_p} \wedge (\frac{\partial B_{k_1...k_q}}{\partial x^i} dx^i \wedge$$

$$\wedge dx^{k_1} \wedge \ ... \ \wedge dx^{k_q}) = dA \wedge B + (-1)^p A \wedge dB$$

$$3\text{-}9)$$

and finally, to create the consistency of rule (d), it is needed just

$$d\left(dA\right)=d\left(\frac{\partial A_{j_1\ldots j_p}}{\partial x^{k}}dx^{k}\wedge dx^{j_1}\wedge\ \ldots\ \wedge dx^{j_p}\right)$$

$$=\frac{\partial^2 A_{j_1\ldots j_p}}{\partial x^{i}\partial x^{k}}dx^{i}\wedge dx^{k}\wedge dx^{j_1}\wedge\ \ldots\ \wedge dx^{j_p}\equiv 0$$

(13-10)

Therefore we conclude that the operation d is, well-defined.

14. Lie bracket and Lie differentiation

Let X and Y be two vector fields and (f) be an arbitrary function. We define their Lie bracket, [X, Y], by its action on f, as following

$$[X, Y]f= (XY- YX)f= X(Yf)- Y(Xf). \qquad (14\text{-}1)$$

Now we want to consider the Lie bracket of any two tangent vectors. Since

$$[X, Y](\alpha f +\beta g)= \alpha[X, Y]f +\beta[X, Y]g \qquad (14\text{-}2)$$

and

$$[X, Y](fg)= g[X, Y]f +f[X, Y]g \qquad (14\text{-}3)$$

where α and β are two real numbers and f and g are two arbitrary functions, so we conclude that, the Lie bracket of any two tangent vectors is a tangent vector. We know, the relation (14-2) is manifest but about the relation (14-3), it follows easily:

$[X, Y](fg) = X(Yfg)- Y(Xfg)$

$\qquad = X(gYf+fYg)- Y(gXf+fXg)$

$$= gXYf+(Xg)(Yf)+(Xf)(Yg)+fXYg$$

$$-\{gYXf+(Xf)(Yg)+(Yf)(Xg)+fYXg\} \qquad (14\text{-}4)$$

$$= g[X,Y]f+f[X,Y]g$$

Therefore the Lie bracket is established as a linear operator in relation (14-2) while the relation (14-3) puts Lie bracket as a differentiation.

We can easily verify that the Lie bracket satisfies the Jacobi identity,

$$[[X, Y], Z] + [[Y, Z], X] + [[Z, X], Y] = 0. \qquad (14\text{-}5)$$

We know that the Lie bracket of X and Y is a tangent vector. In a local coordinate basis, we can obtain its components by its action on x^j. Therefore,

$$[X,Y\,]^j = (XY - YX)x^j = XY^j - YX^j$$

$$= X^k Y^j_{,k} - Y^k X^j_{,k} \qquad (14\text{-}6)$$

where a comma before an index indicates partial differentiation with respect to that index.

It is obvious that the Lie bracket $[\partial_k, \partial_j]$ is zero in a local coordinate basis.

When we consider the Lie bracket as a differentiation, we call [X, Y], the Lie derivative of Y in the X direction, and we write it as:

$$L_X Y = [X,Y\,] = -[Y,X\,] = -L_Y X . \qquad (14\text{-}7)$$

Generally, the Lie derivative, $L_X T$, where T is a tensor field of a given type, can be defined as a same type tensor which obey the following rules:

(a) If it acts on a scalar field f , we will have

$$L_X f = Xf = df (X)$$
(14-8)

and (b) when it acts on a tangent vector Y, as it has been already defined, we have,

$$L_X Y = [X ,Y]$$
(14-9)

(c) On tensor fields $L_X T$, operates linearly, and when acting on tensor products, it satisfies the Liebnitz rule

$$L_X (S \otimes T) = L_X S \otimes T + S \otimes L_X T$$
(14-10)

where T and S are two arbitrary tensor fields.

The rule (c), enables us to calculate the effect of L_X on an arbitrary type tensor. So, by considering the contracted version of the relation

$$L_X (\omega \otimes Y) = (L_X \omega) \otimes Y + \omega \otimes (L_X Y)$$
(14-11)

We can determine its effect on a one form ω for any arbitrary vector-field Y, namely,

$$L_X \langle \omega, Y \rangle = \langle L_X \omega, Y \rangle + \langle \omega, L_X Y \rangle$$
(14-12)

If we write the equation (14-12) obviously, we have

$$X^k(\omega_j Y^{\ j})_{,k} = (L_X \omega)_j Y^{\ j} + \omega_j (L_X Y)^j \qquad (14\text{-}13)$$

Or using equation (14-6), we find,

$$
\begin{aligned}
(L_X \omega)_j Y^{\ j} &= X^k(\omega_{j,k} Y^{\ j} + \omega_j Y^{\ j}_{,k}) - \omega_j(X^k Y^{\ j}_{,k} - Y^k X^{\ j}_{,k}) \\
&= (X^k \omega_{j,k} + \omega_k X^{\ k}_{,j}) Y^{\ j}
\end{aligned}
\qquad (14\text{-}14)
$$

From equation (14-14) (which is valid for an arbitrary Y), we deduce that

$$(L_X \omega)_j = \omega_{j,k} X^k + \omega_k X^{\ k}_{,j} . \qquad (14\text{-}15)$$

The equation (14-12) can be written alternatively in the following form,

$$L_X[\omega(Y)] = (L_X \omega)(Y) + \omega(L_X Y) \qquad (14\text{-}16)$$

Using rule (c), equation (14-16) can be generalized to a type (r,s) tensor, so we can write:

$$
\begin{aligned}
L_X[\mathrm{T}(\omega^1,...,\omega^r,Y_1,...,Y_s)] &= (L_X \mathrm{T})(\omega^1,...,\omega^r,Y_1,...,Y_s) \\
+T(L_X \omega^1, \omega^2,...,\omega^r,Y_1,...,Y_s) &+...+ \mathrm{T}(\omega^1,...,\omega^r,Y_1,...,L_X Y_s)
\end{aligned}
\qquad (14\text{-}17)
$$

where except the first term on the right side of this equation, we can evaluate all other terms, in terms of the known results (14-6), (14-8), and (14-15). Therefore we can deduce the components of L_X from equation (14-17).

Now we want to derive an identity relating the exterior derivative of a one-form to Lie derivatives. So, using equation (14-15), we have

$$\langle L_X \omega,Y \rangle - Y \langle \omega,X \rangle = (\omega_{j,k} X^k + \omega_k X^k_{,j})Y^j - Y^j(\omega_{k,j} X^k + \omega_k X^k_{,j})$$
$$= (\omega_{j,k} + \omega_{k,j})X^k Y^j = 2d\omega(X,Y)$$

$$(14\text{-}18)$$

If we substitute $\langle L_X \omega,Y \rangle$ from equation (14-12), the required result will be obtained as following:

$$d\omega(X,Y) = \frac{1}{2}\{X \langle \omega,Y \rangle - Y \langle \omega,X \rangle - \langle \omega,[X,Y] \rangle\} \qquad (14\text{-}19)$$

where the Lie bracket [X,Y] is written instead of $L_X Y$.

15. Covariant differentiation

We want to define a kind of differentiation which, unlike exterior differentiation, there is another additional structure in that, which its name is affine connection, ∇. The affine connection assigns a differential operator, ∇_X, to each vector field X on M. This operator maps an arbitrary vector-field, Y, into a vector field $\nabla_X Y$.

We put the following conditions which they are consistent with the above requirements,

(a) For any two arbitrary functions f and g which are defined on M, in the argument X , $\nabla_X Y$ is linear, i.e.,

$$\nabla_{fX+gY} Z = f\nabla_X Z + g\nabla_Y Z \qquad (X,Y,Z \in T_0^1) \qquad (15\text{-}1)$$

63

(b) $\qquad \nabla_X (Y + Z) = \nabla_X Y + \nabla_X Z \qquad (X, Y, Z \in T_0^1) \qquad$ (15-2)

(c) $\qquad \nabla_X f = Xf ,\qquad$ (15-3)

(d) $\qquad \nabla_X (fY) = (\nabla_X f)Y + f \nabla_X Y .\qquad$ (15-4)

where f is an arbitrary function on M.

We must pay attention that, in a local coordinate basis (∂_k), from equation (15-3), we deduce that, when ∇_{∂_k} acts on functions, it coincides with partial differentiation with respect to x^k .

Now we want to define the covariant derivative, as a type (1,1) tensor field,

$$\nabla Y (X) = \langle \nabla Y, X \rangle = \nabla_X Y \qquad (15-5)$$

where $X \in T_0^1$. In this notation, equation (15-4) can be rewritten in the following form

$$\nabla(fY) = df \otimes Y + f \nabla Y \qquad (15-6)$$

Rewriting $\nabla_X Y$ relative to some chosen dual bases (e_i) and (e^j), clarify the exact meaning of the assignment of a connection. So, using rules (a)-(d), we have

$$\nabla_X Y = \nabla_X (Y^j e_j) = (XY^j)e_j + Y^j \nabla_X e_j \qquad (15\text{-}7)$$

Since for a particular e_j, $\nabla_X e_j$ is a type (1,0) tensor field, in the chosen basis, we should have a representation of the form

$$\nabla_X e_j = \omega_j^l(X)e_l \qquad (15\text{-}8)$$

Where depending on l and j, ω_j^l are one-forms. Consequently, we write

$$\nabla_X Y = (XY^j)e_j + Y^j \omega_j^l(X)e_l \qquad (15\text{-}9)$$

and equation (15-7) can be written in the following form,

$$\begin{aligned}
\nabla_X Y &= (XY^j)e_j + Y^j \nabla_{X^k e_k} e_j \\
&= (XY^j)e_j + Y^j X^k \nabla_{e_k} e_j
\end{aligned} \qquad (15\text{-}10)$$

or correspond to equation (15-8),

$$\nabla_X Y = (XY^j)e_j + Y^j X^k \omega_j^l(e_k)e_l \qquad (15\text{-}11)$$

In the basis (e^k), if we suppose that in the expansion of ω_j^l,

$$\omega_j^l(e_k) = \omega_{jk}^l \qquad (15\text{-}12)$$

be the coefficient of e^k, we will deduce that a connection ∇ is specified to ω^l_{jk}, by n^2 one-forms ω^l_j or, equivalently, n^3 scalar fields.

If we write the equation (15-9) in the following form,

$$\nabla_X Y = [XY^{\;j} + \omega^j_l (X) Y^{\;l}] e_j \qquad (15\text{-}13)$$

we conclude that,

$$(\nabla_X Y)^j = XY^{\;j} + \omega^j_l (X) Y^{\;l} \qquad (15\text{-}14)$$

Using equation (15-14), in a local coordinate basis (∂_k, dx^l), we can write

$$(\nabla_{\partial_k} Y)^j = \partial_k Y^{\;j} + Y^{\;l} \omega^j_{lk} = Y^{\;j}_{,k} + Y^{\;l} \omega^j_{lk} \qquad (15\text{-}15)$$

Now if we write Γ^j_{lk} in place of ω^j_{lk} and use semicolons for indicating covariant derivatives (in contrast to comma that used to indicate ordinary derivatives), we find the following formula:

$$Y^{\;j}_{;k} = Y^{\;j}_{,k} + Y^{\;l} \Gamma^j_{lk} \qquad (15\text{-}16)$$

If the operation of ∇ satisfies the Leibnitz rule when acting on tensor products, then we can extend the definition of covariant derivatives of vector fields, to tensor fields. So it is required that,

$$\nabla(S \otimes T) = \nabla S \otimes T + S \otimes \nabla T \qquad (15\text{-}17)$$

where T and S are two tensor fields. The outcome of the above requirement is (compare with equation (14-17))

$$\nabla_X \{T(\omega^1, \ldots, \omega^r, Y_1, \ldots, Y_s)\} = (\nabla_X T)(\omega^1, \ldots, \omega^r, Y_1, \ldots, Y_s)$$

$$+T(\nabla_X \omega^1, \omega^2, \ldots, \omega^r, Y_1, \ldots, Y_s) + \ldots$$

$$+ T(\omega^1, \ldots, \omega^r, Y_1, \ldots, Y_{s-1}, \nabla_X Y_s) \qquad (15\text{-}18)$$

From the above equations we conclude that, for every vector field Y, if Ω is a one-form, then we can write,

$$\nabla_X (\Omega(Y)) = (\nabla_X \Omega)(Y) + \Omega(\nabla_X Y) \qquad (15\text{-}19)$$

If we write the equation (15-19) in terms of local basis (e^j) and (e_i), then we will have,

$$\nabla_X (\Omega_j (Y^j)) = (\nabla_X \Omega)_j Y^j + \Omega_j (\nabla_X Y)^j \qquad (15\text{-}20)$$

Now using equation (15-14) and rule (c), we have

$$(\nabla_X \Omega)_j Y^j = (X\Omega_j)Y^j + \Omega_j (XY^j) - \Omega_j [XY^j + Y^l \omega_l^j (X)]$$
$$= (X\Omega_j)Y^j - \Omega_l \omega_j^l (X)Y^j \qquad (15\text{-}21)$$

We infer that,

$$(\nabla_X \Omega)_j = X\Omega_j - \Omega_l \omega_j^l (X) \qquad (15\text{-}22)$$

Or we have

$$\nabla_X \Omega = [X\Omega_j - \Omega_l \omega_j^l (X)]e^j \qquad (15\text{-}23)$$

If in equation (15-23) we put $\Omega = e^j$, then we obtain the following formula

$$\nabla_X e^j = -\omega_l^j (X)]e^l \qquad (15\text{-}24)$$

that we can compare it with the formula (15-8).

Equation (15-24) shows that, once we accept the Leibnitz rule for tensor products, to determine the covariant derivatives of one-forms, a knowledge of the n^2 one-forms ω_j^l are sufficient. Also from equation (15-22) we have:

$$\Omega_{j;k} = \Omega_{j,k} - \Omega_l \Gamma_{jk}^l \qquad (15\text{-}25)$$

If we apply equations (15-22) and (15-25) to the one-form df, then we will get an important result.

We know, in a local coordinate basis the components of df are $f_{,j}$, so using equation (15-25) we can write:

$$f_{,j;k} = f_{,j,k} - f_{,l} \Gamma_{jk}^l \qquad (15\text{-}26)$$

Now, if we permute the indices k and j in equation (15-25), we find

$$f_{,k;j} = f_{,k,j} - f_{,l} \Gamma_{kj}^l \qquad (15\text{-}27)$$

Paying attention to equation (15-18), we see that, using equations (15-13) and (15-22), the covariant derivative of an arbitrary tensor-field can be written as following:

$$f_{,j;k} - f_{,k;j} = -f_{,l} (\Gamma_{jk}^l - \Gamma_{kj}^l) \qquad (15\text{-}28)$$

16. Space-time as a five-dimensional manifold

In Electrogravity theory, we consider the space-time as a five dimensional differentiable manifold.

In a particular space time, we can write the following Minkowskian like metric:

$$ds^2 = c^2 dt^2 - dx_1^2 - dx_2^2 - dx_3^2 - dx_4^2 \qquad (16\text{-}1)$$

where c is the velocity of light (which by a special choice of units we set it equal to 1).

Since in above metric with its lorentzian signature, the $g(X,X)$ can be positive, zero, or negative, so the space-time metric is not positive-definite. If $g(X,X) > 0$, $= 0$, or < 0, then we call vectors accordingly, as time-like, null, or space-like.

Time-like paths, which are curves that the tangent vector is always time-like along them, can be described by material particles with finite rest mass. But null trajectories which are curves that the tangent vector is always null along them, can be described by massless particles (such as photons).

If the freely falling particles have finite rest mass, they will describe the time-like geodesics and if they are massless, they will describe null geodesics. Even though the null geodesies can be affinely parametrized by a change of variables, but we cannot parameterize them clearly, by arc length.

We know, the Ricci identity has the form

$$R^i_{jkl}Z_i = Z_{j;k;l} - Z_{j;l;k} \tag{16-2}$$

We can obtain the Ricci tensor by contraction

$$g^{jl}R_{ijkl} = R_{ik} \tag{16-3}$$

and the Weyl tensor is as following

$$C_{ijkl} = R_{ijkl} - \frac{1}{2}(g_{ik}R_{jl} + g_{jl}R_{ik} - g_{jk}R_{il} - g_{il}R_{jk})$$
$$+ \frac{1}{6}(g_{ik}g_{jl} + g_{jk}g_{il})R \tag{16-4}$$

17. The pentad formalism

In Electrogravity theory, the standard way of handling problems that comes in to mind, is to consider the field-equations in a local coordinate basis adapted to the special problem. But it is appeared that, there are some advantageous to treat in a different way, as first we choose a suitable pentad basis of five linearly independent vector-fields, then on to the chosen basis, we project the relevant quantities and then we consider the equations satisfied by them. This is the pentad formalism.

In pentad formalism applications, choosing the pentad basis depends on the space-time symmetries that we want to understand, and sometimes, it is a part of the problem. In addition, always we don't know the relevant equations and the probable relations between them.

Thus, without any prior commitments, we must present the basic ideas of the theory and derive the related equations of the formalism.

18. The pentad representation

We can set up a basis of five contravariant vectors at each point of the space-time,

$$e^i_{(a)} \qquad (a = 1, 2, 3, 4, 5) \qquad (18\text{-}1)$$

Where a in parentheses distinguishes the pentad indices (a), from the tensor indices (i).

There are following covariant vectors associated with the above contravariant vectors in (18-1)

$$e_{(a)i} = g_{ik} e^k_{(a)} \qquad (18\text{-}2)$$

where g_{ik} is the metric tensor. Besides, we can define the inverse of the matrix $[e^i_{(a)}]$ as $e^{(b)}_i$, so that

$$e^i_{(a)} e^{(b)}_i = \delta^{(b)}_{(a)} \qquad \text{and} \qquad e^i_{(a)} e^{(a)}_j = \delta^i_j \qquad (18\text{-}3)$$

Now, also we can assume that,

$$e^i_{(a)} e_{(b)i} = \eta_{(a)(b)} \qquad (18\text{-}4)$$

where $\eta_{(a)(b)}$ is a constant symmetric matrix.

The basis vectors, $e^i_{(a)}$, are usually supposed to be orthonormal in which case $\eta_{(a)(b)}$ is a diagonal matrix and its diagonal elements are, $+1, -1, -1, -1, -1$.

We can develop a more general formalism, in which the $\eta_{(a)(b)}$'s are functions on the manifold. But for present, we assume that the $\eta_{(a)(b)}$'s are constants.

We suppose that the $\eta^{(a)(b)}$ is the inverse of the matrix $[\eta_{(a)(b)}]$, thus,

$$\eta^{(a)(b)}\eta_{(b)(c)} = \delta^{(a)}_{(c)} \qquad (18\text{-}5)$$

From the foregoing definitions, we can write,

$$\eta_{(a)(b)}e^{(a)}_i = e_{(b)i}, \qquad \eta^{(a)(b)}e_{(a)i} = e^{(b)}_i \qquad (18\text{-}6)$$

and,

$$e_{(a)i}e^{(a)}_j = g_{ij} \qquad (18\text{-}7)$$

For obtaining the pentad components of an arbitrary vector or tensor field, it must be projected onto pentad frame. So we have,

$$\left.\begin{array}{ll} A_{(a)} = e_{(a)j}A^j & = e^j_{(a)}A_j \\ A^{(a)} = \eta^{(a)(b)}A_{(b)} = e^{(a)}_j A^j = e^{(a)j}A_j \\ A^i = e^i_{(a)}A^{(a)} & = e^{(a)i}A_{(a)} \end{array}\right\} \qquad (18\text{-}8)$$

and we can write more generally,

$$\left.\begin{array}{l} T_{(a)(b)} = e^{i}_{(a)} e^{j}_{(b)} T_{ij} = e^{i}_{(a)} T_{i\,(b)} \\ T_{ij} = e^{(a)}_{i} e^{(b)}_{j} T_{(a)(b)} = e^{(a)}_{i} T_{(a)j} \end{array}\right\} \qquad (18\text{-}9)$$

From equations (18-3), (18-4), (18-5), and (18-6) we can conclude that,

a) It can be passed from pentad indices to the tensor indices and vice versa.

b) Using $\eta_{(a)(b)}$ and $\eta^{(a)(b)}$ we can raise and lower the pentad indices, even as using the metric tensor we can raise and lower the tensor indices.

c) There isn't any ambiguity in having quantities with pentad indices or tensor indices.

d) If we contract a tensor with respect to its tensor or pentad indices, the result of contraction is the same.

19. Directional derivative

We can define the directional derivatives using contravariant vectors $e_{(a)}$, that are considered as tangent vectors,

$$e^{(a)} = e^{i}_{(a)} \frac{\partial}{\partial x^{i}} \qquad (19\text{-}1)$$

and we write

$$\phi_{,(a)} = e^{i}_{(a)} \frac{\partial \phi}{\partial x^{i}} = e^{i}_{(a)} \phi_{,i} \qquad (19\text{-}2)$$

where ϕ is a scalar field. Now we define generally,

73

$$A_{(a),(b)} = e^i_{(b)} \frac{\partial}{\partial x^i} A_{(a)} = e^i_{(b)} \frac{\partial}{\partial x^i} e^j_{(a)} A_j$$

$$= e^i_{(b)} \nabla_{\partial_i} [e^j_{(a)} A_j] = e^i_{(b)} [e^j_{(a)} A_{j;i} + A_k e^k_{(a);i}] \tag{19-3}$$

So we obtain

$$A_{(a),(b)} = e^j_{(a)} A_{j;i} e^i_{(b)} + e_{(a)k;i} e^i_{(b)} e^k_{(c)} A^{(c)} \tag{19-4}$$

Using the following definition

$$\gamma_{(c)(a)(b)} = e^k_{(c)} e_{(a)k;i} e^i_{(b)} \tag{19-5}$$

equation (19-4) can be written as

$$A_{(a),(b)} = e^j_{(a)} A_{j;i} e^i_{(b)} + \gamma_{(c)(a)(b)} A^{(c)} \tag{19-6}$$

The $\gamma_{(c)(a)(b)}$ is called the Ricci rotation-coefficients which is defined in equation (19-5). We can define these coefficients equivalently as following

$$e_{(a)k;i} = e^{(c)}_k \gamma_{(c)(a)(b)} e^{(b)}_i \tag{19-7}$$

These coefficients are antisymmetric in the first pair of indices

$$\gamma_{(c)(a)(b)} = -\gamma_{(a)(c)(b)} \tag{19-8}$$

If we expand the identity, we reach the following fact,

$$0 = \eta_{(a)(b),i} = [e_{(a)k} e^k_{(b)}]_{;i} \tag{19-9}$$

We must pay attention that, if $\eta_{(a)(b)}$'s had not been constant, we couldn't have inferred this antisymmetry. Due to this antisymmetry, we can write equation (19-7) as

$$e^{k}_{(a);i} = -\gamma^{k}_{(a)i} \tag{19-10}$$

Now we can write the equation (19-6), in the following alternative form

$$e^{i}_{(a)}A_{i;j}e^{j}_{(b)} = A_{(a),(b)} - \eta^{(n)(m)}\gamma_{(n)(a)(b)}A_{(m)} \tag{19-11}$$

We call the intrinsic derivative of $A_{(a)}$ in the direction of $e_{(b)}$, to the quantity on the right side of the equation (19-11), and write it as $A_{(a),(b)}$:

$$A_{(a)|(b)} = e^{i}_{(a)}A_{i;j}e^{j}_{(b)} \qquad (\text{or, } A_{i;j} = e^{(a)}_{i}A_{(a)|(b)}e^{(b)}_{j}) \tag{19-12}$$

Thus, relating the intrinsic and directional derivatives, we have the following formula,

$$A_{(a)|(b)} = A_{(a),(b)} - \eta^{(n)(m)}\gamma_{(n)(a)(b)}A_{(m)} \tag{19-13}$$

Therefore, from (19-12) we conclude that, it can be freely passed from covariant derivatives to intrinsic derivatives and vice versa.

We can readily extend the intrinsic derivative of vector fields to tensor fields. So for Riemann tensor, we can write the intrinsic derivative of it as following,

$$R_{(a)(b)(c)(d)|(f)} = R_{ijkl;m}e^{i}_{(a)}e^{j}_{(b)}e^{k}_{(c)}e^{l}_{(d)}e^{m}_{(f)} \tag{19-14}$$

and then,

$$R_{(a)(b)(c)(d),(f)} = [R_{ijkl}e^i_{(a)}e^j_{(b)}e^k_{(c)}e^l_{(d)}]_{;m}e^m_{(f)} \qquad (19\text{-}15)$$

Now according the relation (19-10), if we replace the respective rotation-coefficients, in place of the covariant derivatives of the different basis-vectors, we will find

$$R_{(a)(b)(c)(d)|(f)} = R_{(a)(b)(c)(d),f} - \eta^{(n)(m)}[\gamma_{(n)(a)(f)}R_{(m)(b)(c)(d)} + \gamma_{(n)(b)(f)}R_{(a)(m)(c)(d)}$$
$$+ \gamma_{(n)(c)(f)}R_{(a)(b)(m)(d)} + \gamma_{(n)(d)(f)}R_{(a)(b)(c)(m)}]$$
$$(19\text{-}16)$$

Now we observe that, for evaluating the rotation coefficients we don't require to evaluate the covariant derivatives and the Christoffel symbols.

If we define,

$$\lambda_{(a)(b)(c)} = e_{(b)i,j}[e^i_{(a)}e^j_{(c)} - e^j_{(a)}e^i_{(c)}] \qquad (19\text{-}17)$$

and rewrite it, in the following form,

$$\lambda_{(a)(b)(c)} = [e_{(b)i,j} - e_{(b)j,i}]e^i_{(a)}e^j_{(c)} \qquad (19\text{-}18)$$

we will see that, the ordinary derivatives of $e_{(b)i}$ and $e_{(b)j}$ can be replaced by the corresponding covariant derivatives

$$\lambda_{(a)(b)(c)} = [e_{(b)i;j} - e_{(b)j;i}]e^i_{(a)}e^j_{(c)}$$
$$= \gamma_{(a)(b)(c)} - \gamma_{(c)(b)(a)} \qquad (19\text{-}19)$$

Using this relation we have

$$\gamma_{(a)(b)(c)} = \frac{1}{2}[\lambda_{(a)(b)(c)} + \lambda_{(c)(a)(b)} + \lambda_{(b)(c)(a)}] \qquad (19\text{-}20)$$

and using equation (19-18), we see that, for evaluating the $\lambda_{(a)(b)(c)}$, we only require the evaluation of ordinary derivatives. Here, we must pay attention that the λ-symbols are antisymmetric in the first and the third indices:

$$\lambda_{(a)(b)(c)} = -\lambda_{(c)(b)(a)} \qquad (19\text{-}21)$$

20- Postulates

In the new theory, we need to modify the light speed constancy postulate. Let K be a system of reference in which a mass distant from other masses is moving with uniform motion on a straight line and K' be another system of reference which is moving relatively to K in accelerated translation. If a light ray propagates in a straight line with a constant velocity with respect to K, the path of light will be curvilinear with respect to K'.

But about the other postulate, we accept the general covariance principle here, which is, the general laws of nature are to be expressed by equations

which hold good for all systems of coordinates.

21- Occurrence of the Electrogravity Field in a Space-time

In this book, we are presenting the Electrogravity (EG) theory for unifying gravitation with electromagnetism in a five dimensional space-time.

If we choose a system of coordinate in the finite region in such a way that the $g_{\mu\nu}$ has constant values, we will see that a free material point moves, relative to this system, with uniform motion on a straight line. But if we choose a new space-time coordinates x_1, x_2, x_3, x_4, and $x_5,$, the $g_{\mu\nu}$ in the new system will not be constant, but functions of space and time, and the motion of the free material point will be a curvilinear motion. This motion should be interpreted as a motion under the influence of the Electrogravity field. So we find the occurrence of an EG field connected with the space-time variables of $g_{\mu\nu}$. So the $g_{\mu\nu}$ representing the EG field at the same time define the metrical properties of the space-time.

Now we want to define a five vector as following[1],

In a $(x_1, x_2, x_3, x_4, x_5)$ coordinate, using the following metric,

$$ds^2 = -A_1^2 dx_1^2 - A_2^2 dx_2^2 - A_3^2 dx_3^2 - A_4^2 dx_4^2 + A_5^2 dx_5^2 \qquad (21\text{-}1)$$

we define a five vector as following:

$$h_\mu = h_1 \hat{e}^1 + h_2 \hat{e}^2 + h_3 \hat{e}^3 + h_4 \hat{e}^4 + h_5 \hat{e}^5 \qquad (21\text{-}2)$$

---

[1] book: Gravitation"by Charles W.Misner, Kip S.Thorne, and John Archibald Wheeler,chapter 3. and
book: A first course in general relativity" by Bernard F.Schutz, Cambridge university, chapter 3.

78

where $\hat{e}^i$ are the unit vectors on x_i direction and:

$$h_1 = A_1, h_2 = A_2, h_3 = A_3,\ h_4 = A_4\ ,\ h_5 = A_5 \qquad (21\text{-}3)$$

Since, $$g^{\alpha\tau} R^{\mu}_{\upsilon\sigma\tau} = R^{\mu\alpha}_{\upsilon\sigma} \qquad (21\text{-}4)$$

where $R^{\mu}_{\upsilon\sigma\tau}$ is the Riemann tensor, one can write:

$$h_\alpha R^{\mu\alpha}_{\upsilon\sigma} = R^{\mu\alpha}_{\upsilon\sigma\alpha} \qquad (21\text{-}5)$$

If we contract (21-5) two times, we will obtain:

$$R^{\mu\alpha}_{\upsilon\mu\alpha} = R_\upsilon \qquad (21\text{-}6)$$

Now to find the geometrical form of the antisymmetric tensor, which was discussed in introduction, and find the relationship between this tensor and gravitational geometry, we define $S_{\mu\upsilon}$ using the following wedge product:

$$S_{\mu\upsilon} = R_\mu \wedge h_\upsilon \qquad (21\text{-}7)$$

where $S_{\mu\upsilon}$ is an antisymmetric tensor. Now we define the EG tensor as:

$$\frac{\partial S_{\mu\upsilon}}{\partial x_\sigma} - \Gamma^\tau_{\mu\sigma} S_{\tau\upsilon} - \Gamma^\tau_{\sigma\upsilon} S_{\mu\tau} + h_\mu R_{\upsilon\sigma} \qquad (21\text{-}8)$$

where $R_{\upsilon\sigma}$ is the Ricci tensor.

22. The Geodesic Equation

In order to find the geodesic equation, we use the variational principle.

For a free material point motion between two fixed points in a space-time we have: $L = T - U$.   Where T and U are kinetic and potential energies and L is Lagrangian. We know, for a free material point motion:

$$U=0 \text{ and } T = g_{\alpha\beta}\dot{x}^{\alpha}\dot{x}^{\beta} \qquad (22\text{-}1)$$

Since we have: $\delta \int_{p}^{p'} ds = 0$, using Lagrange-Euler equation:

$$-\frac{\partial L}{\partial q} + \frac{d}{dt}\frac{\partial L}{\partial \dot{q}} = 0 \quad \Rightarrow \quad \frac{d}{d\lambda}(\frac{\partial L}{\partial \dot{x}^{\alpha}}) - \frac{\partial L}{\partial x^{\alpha}} = 0 \qquad (22\text{-}2)$$

we have:

$$L_{,\alpha} = \frac{\partial L}{\partial x^{\alpha}} = g_{\mu\upsilon,\alpha}\dot{x}^{\mu}\dot{x}^{\upsilon} \quad , \quad \frac{\partial L}{\partial \dot{x}^{\alpha}} = g_{\alpha\upsilon}\dot{x}^{\upsilon} + g_{\mu\alpha}\dot{x}^{\mu} \qquad (22\text{-}3)$$

and,

$$\frac{d}{d\lambda}(\frac{\partial L}{\partial \dot{x}^{\alpha}}) = \frac{d}{d\lambda}(g_{\alpha\upsilon}\dot{x}^{\upsilon} + g_{\mu\alpha}\dot{x}^{\mu}) = g_{\alpha\upsilon,\mu}\dot{x}^{\mu}\dot{x}^{\upsilon} + g_{\alpha\upsilon}\ddot{x}^{\upsilon} + g_{\mu\alpha,\upsilon}\dot{x}^{\mu}\dot{x}^{\upsilon} + g_{\mu\alpha}\ddot{x}^{\mu}$$

$$(22\text{-}4)$$

from (22-2), (22-3) and (22-4) we have:

$$g_{\alpha\upsilon}\ddot{x}^{\upsilon} + g_{\mu\alpha}\ddot{x}^{\mu} + \dot{x}^{\upsilon}\dot{x}^{\mu}(g_{\alpha\upsilon,\mu} + g_{\mu\alpha,\upsilon} - g_{\mu\upsilon,\alpha}) = 0 \quad \text{and,}$$

$$g_{\alpha\upsilon}\ddot{x}^{\upsilon} + g_{\mu\alpha}\ddot{x}^{\mu} = \ddot{x}_{\alpha} + \ddot{x}_{\alpha} = 2\ddot{x}_{\alpha} = 2g_{\mu\alpha}\ddot{x}^{\mu} \qquad (22\text{-}5)$$

therefore: $\quad 2g_{\mu\alpha}\ddot{x}^{\mu}+\dot{x}^{\upsilon}\dot{x}^{\mu}(g_{\alpha\upsilon,\mu}+g_{\mu\alpha,\upsilon}-g_{\mu\upsilon,\alpha})=0 \qquad\qquad (22\text{-}6)$

Multiplying both sides of this equation by $\dfrac{1}{2}g^{\mu\alpha}$, we get:

$$\ddot{x}^{\mu}+\frac{1}{2}g^{\mu\alpha}(g_{\alpha\upsilon,\mu}+g_{\mu\alpha,\upsilon}-g_{\mu\upsilon,\alpha})\dot{x}^{\upsilon}\dot{x}^{\mu}=0 \qquad\qquad (22\text{-}7)$$

Since, $\Gamma^{\mu}_{\mu\upsilon}=\dfrac{1}{2}g^{\mu\alpha}(g_{\alpha\upsilon,\mu}+g_{\mu\alpha,\upsilon}-g_{\mu\upsilon,\alpha})$ then equation (22-7) can be written as:

$$\ddot{x}^{\mu}+\Gamma^{\mu}_{\mu\upsilon}\dot{x}^{\upsilon}\dot{x}^{\mu}=0 \qquad\qquad (22\text{-}8)$$

which is the geodesic equation.

23. The Electrogravity Field Equations in the Absence of Matter

Here, we denote "matter", to everything but the gravitational and electromagnetic fields. Then try to find the Electrogravity field equations in the absence of matter.

There is a mathematical importance for the EG tensor which is, if we choose a coordinate system that the $g_{\mu\nu}$ are constant with reference to it, then all components of the EG tensor will vanish. If we choose any new system of coordinates, instead of the original one, the $g_{\mu\nu}$ are not constant in the new coordinate, but in result of its tensor nature, the transformed components of the EG tensor will still vanish in the new system. Relative to this system, all components of the EG tensor vanish in any other system of coordinates. Therefore, the equations that are required for the matter-free EG field must be satisfied if all components of the EG tensor vanish.

Thus using eq.(21-8), the equations of matter-free field are:

$$\frac{\partial S_{\mu\nu}}{\partial x_{\sigma}} - \Gamma^{\tau}_{\mu\sigma}S_{\tau\nu} - \Gamma^{\tau}_{\sigma\nu}S_{\mu\tau} + h_{\mu}R_{\nu\sigma} = 0 \tag{23-1}$$

In approximation for considering the electromagnetic behaviors of this equation, it is better to focus on the microscopic scales. If some electric charges are placed in the media, it is obvious that, in this scale the effects of gravitational field are so much smaller than electromagnetic field, so the gravitational field can be ignored in approximation. On the other hand, since the existing mass (the mass of the electric charges) is very small, then the energy momentum tensor ($T_{\nu\sigma}$) is very small and therefore $R_{\nu\sigma}$ will be very small and ignorable in this scale, and expression (21-8) turns into:

$$\frac{\partial S_{\mu\nu}}{\partial x_{\sigma}} - \Gamma^{\tau}_{\mu\sigma}S_{\tau\nu} - \Gamma^{\tau}_{\sigma\nu}S_{\mu\tau} \tag{23-2}$$

Using (23-2) , the following two tensors can be written additionally:

$$\frac{\partial S_{\nu\sigma}}{\partial x_{\mu}} - \Gamma^{\tau}_{\mu\nu}S_{\tau\sigma} - \Gamma^{\tau}_{\mu\sigma}S_{\nu\tau} \tag{23-3}$$

$$\frac{\partial S_{\sigma\mu}}{\partial x_{\nu}} - \Gamma^{\tau}_{\nu\sigma}S_{\tau\mu} - \Gamma^{\tau}_{\nu\mu}S_{\sigma\tau} \tag{23-4}$$

In eq.(23-1), it is seen that all components of the EG tensor vanished, thus the sum of the above three tensors ((23-2), (23-3) and (23-4)) must be zero also. Now adding (23-2), (23-3) and (23-4), it will be obtained:

$$\frac{\partial S_{\mu\upsilon}}{\partial x_\sigma} + \frac{\partial S_{\upsilon\sigma}}{\partial x_\mu} + \frac{\partial S_{\sigma\mu}}{\partial x_\upsilon} - \Gamma^\tau_{\mu\sigma} S_{\tau\upsilon} - \Gamma^\tau_{\sigma\upsilon} S_{\mu\tau} - \Gamma^\tau_{\mu\upsilon} S_{\tau\sigma} -$$
$$\Gamma^\tau_{\mu\sigma} S_{\upsilon\tau} - \Gamma^\tau_{\upsilon\sigma} S_{\tau\mu} - \Gamma^\tau_{\upsilon\mu} S_{\sigma\tau} = 0 \qquad (23\text{-}5)$$

Using $S_{\mu\upsilon} = -S_{\upsilon\mu}$ in equation (23-5) and simplifying it, we get:

$$\frac{\partial S_{\mu\upsilon}}{\partial x_\sigma} + \frac{\partial S_{\upsilon\sigma}}{\partial x_\mu} + \frac{\partial S_{\sigma\mu}}{\partial x_\upsilon} = 0 \qquad (23\text{-}6)$$

We will come back to this formula later.

24. Equation of Motion of a Material Point in the Electrogravity Field

In special theory of relativity in the absence of the external forces, a freely movable body moves uniformly on a straight line. If we choose a system of coordinate, K_0, in a five dimensional space-time in such a way that the $g_{\mu\nu}$ s are constant, then the material point will be moved on a straight line. If this movement is considered from any other system of coordinates, K_1, the law of motion with respect to K_1 results from the following consideration. The movement of body in K_0 coordinate corresponds to a five dimensional straight line which is a geodetic line. Since the geodetic line is independent from the system of reference, its equations will also be the equation of motion of the material point with respect to K_1. So the equations which define the motion of the point with respect to K_1 is:

$$\ddot{x}^\upsilon + \Gamma^\upsilon_{\sigma\tau}\dot{x}^\sigma\dot{x}^\tau = 0 \qquad (24\text{-}1)$$

Now we say that, the above equation also defines the movement of a particle in the EG field.

25. The General Form of the Electrogravity Field Equations

The field equations (23-1), which are obtained for matter free space-time, are to be compared with the field equation $\nabla^2\varphi = 0$ of Newton's theory or with $R_{\mu\upsilon} = 0$ of Einstein's gravity field equations in vacuum. We need an equation corresponding to Poisson's equation: $\nabla^2\varphi = 4k\,\pi\rho$ or $R_{\mu\upsilon} - \dfrac{1}{2}g_{\mu\upsilon}R = -kT_{\mu\upsilon}$ of Einstein's general form of the gravitational field equation. For this purpose we define $T_{\mu\upsilon\sigma}$ tensor using the combination of electromagnetic energy tensor and energy-momentum tensor:

$$T_{\mu\upsilon\sigma} = g_{\alpha\upsilon}(kT^\alpha_\sigma + k\,T'^\alpha_\sigma)h_\mu \qquad (25\text{-}1)$$

where k and k' are two constants related to the gravity and electromagnetism respectively, and $g_{\alpha\upsilon}T^\alpha_\sigma = T_{\upsilon\sigma}$ is the energy-momentum tensor and $g_{\alpha\upsilon}T'^\alpha_\sigma = T'_{\upsilon\sigma}$ is the electromagnetic energy tensor where:

$$g_{\alpha\upsilon}T'^{\alpha}_{\ \sigma}\ h_{\mu} = h_{\mu}T'_{\upsilon\sigma} = T'_{\mu\upsilon\sigma} \tag{25-2}$$

So instead of eq.(23-1) we write:

$$\frac{\partial S_{\mu\upsilon}}{\partial x_{\sigma}} - \Gamma^{\tau}_{\mu\sigma}S_{\tau\upsilon} - \Gamma^{\tau}_{\upsilon\sigma}S_{\mu\tau} + h_{\mu}R_{\upsilon\sigma} - \frac{1}{2}g_{\upsilon\sigma}h_{\mu}R = -T_{\mu\upsilon\sigma} \tag{25-3}$$

Therefore equation (25-3) is the required general form of the Electrogravity field equations.

This equation contains everything in electromagnetism and gravitation, from Newton's formula for gravitation till Maxwell equations, and in its special forms easily gives us all Maxwell's equations and the general form of the Einstein field equations in general relativity.

Since Einstein has used "the Christoffel symbols" as the field components of gravitation in his general relativity paper[1], so by approximation, in the absence of matter, when the gravitational field is very small: $\Gamma^{\tau}_{\sigma\upsilon} \to 0$, and according to the Einstein field equation in vacuum, $R_{\sigma\upsilon}$ is very small. For example it can be reached to this approximation in a region without any matter unless some electrical charges, like electron, which they have very small masses. Thus in the mentioned approximation, eq.(25-3) turns into:

$$\frac{\partial S_{\mu\upsilon}}{\partial x_{\sigma}} = -T_{\mu\upsilon\sigma} \tag{25-4}$$

---

[1] A. Einstein, The foundation of the General theory of relativity, Doc.30,p179, 1914.

Suppose that, we are working in a area that the current density is j_i and the total electric charge density of the area is ρ_q. Now in one special case, let's suppose that all components of $T_{\mu\nu\sigma}$ are zero unless five components of $T_{5\nu5}$. Now we write $T_{5\nu5}$ as following:

$$T_{5\nu5} \longrightarrow j_\nu = k(\rho_q, j_i) \quad \text{where} \quad i = 1,2,3,4 \tag{25-5}$$

In above approximation, if we ignore the gravitation (because, here we are working in the microscopic scale and in this scale the intensity of electromagnetic field is so much bigger than gravitational field), in the absence of gravity the space time can be considered as the four dimensional part of R_5, and in this space-time, j_ν turns into:

$$j_\nu = (\rho_q, j_1, j_2, j_3). \tag{25-6}$$

Now from eqs.(25-4) and (25-5):

$$\frac{\partial S_{\mu\nu}}{\partial x_\mu} = -k(\rho_q, j_1, j_2, j_3) = j_\nu \Rightarrow \frac{\partial S_{\mu\nu}}{\partial x_\mu} = j_\nu \tag{25-7}$$

where $S_{\mu\nu}$ is an antisymmetric tensor.

Therefore, in equations (23-6) and (25-7), on identifying $S_{\mu\nu}$ with the electromagnetic tensor $F_{\mu\nu}$, one will recognize eqs.(23-6) and (25-7) as the Maxwell's equations. Therefore, $S_{\mu\nu}$ can be identified as the electromagnetic tensor ($F_{\mu\nu}$).

Moreover, equation (25-3) in its special form, easily gives us the general form of the Einstein field equations for gravitation, in the absence of electromagnetic field.

Now, for completing the equations, I think it is better to add a new constant (Λ) to the EG field equations, similar to the thing that Einstein did, which is different from gravitational cosmological constant (Λ_G), and it is considered as the sum of two electromagnetic cosmological constant (Λ_B) and gravitational cosmological constant.

So the complete form of the field equations will be:

$$\frac{\partial S_{\mu\nu}}{\partial x_\sigma} - \Gamma^\tau_{\mu\sigma} S_{\tau\nu} - \Gamma^\tau_{\nu\sigma} S_{\mu\tau} + h_\mu R_{\nu\sigma} - \frac{1}{2} g_{\nu\sigma} h_\mu R + g_{\nu\sigma} h_\mu \Lambda = -T_{\mu\nu\sigma} \quad (25\text{-}8)$$

26. Solution of the Field Equations:

If one tries to find a complete solution for equation (23-1) which contains the effects of both gravity and electromagnetism on space time in a five dimensional coordinate, first he should try to combine two spherical and cylindrical coordinates to make a five dimensional coordinate system, then write the metric in this coordinate and find the metric functions. However one can propose any other logical arbitrary five dimensional coordinate to solve the EG field equations.

After presentation of the Electrogravity field equations, the first people who proposed a solution for these equations, were Shaheen

Ghayourmanesh and Eshagh Soufi. We call to their metric, the Shaheen-Sufi metric[1].

In this book, we just report their solution and we don't want to discuss on the accuracy of their solution here, but we see to it just as an example of a five dimensional solution for E.G field equations. May be, someone like to examine the ellipsoid coordinate for this purpose, or may be another one finds a more complete metric than these, in future.

The general form of the Shaheen-Sufi metric is as following:

$$\mathrm{ds}^2 = A\,\mathrm{dt}^2 - B\,\mathrm{dr}^2 - Cd\theta^2 - Dd\phi^2 - Edz^2 \qquad (26\text{-}1)$$

where the coordinate system is chosen so that: $\quad x = [t, r, \theta, \varphi, z]$.

In Shaheen–Soufi solution, they proposed a 5 dimensional coordinate system, which is obtained from the combination of two spherical and cylindrical coordinates, in which a 4-dimensional spatial coordinate is defined. Then adding the time dimension, they reached a 5-dimensional space-time coordinate system.

27. Four dimensional (4D) coordinate system

As an example, they used the following 4D space coordinate system. Let x_1, x_2, x_3 and x_4 be the space coordinates. This coordinate system is defined using the combination of two spherical and cylindrical coordinates. In this coordinate, both electromagnetic and gravitational

[1] S. Ghayourmanesh ; E. Soufi." Electrogravity field of a charged mass object", International Journal of Science and Research, vol.12, 2023.

curvatures of the space–time, exist simultaneously. They call to this newly defined 4D space coordinate system, the "cylinder-sphere" coordinate, where:

$$\begin{cases} x_1 = r \\ x_2 = \theta \\ x_3 = \varphi \\ x_4 = z \end{cases} \qquad (27\text{-}1)$$

and with $x_5 = t$, we have a five dimensional space-time coordinate system. They defined r, θ, φ and z, using the following figure,

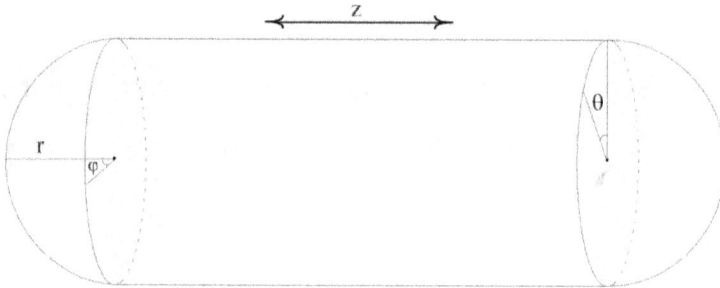

Figure (1)

Where the constraints of each component on the surface of the cylinder-sphere are:

$$\begin{cases} r = 0 \to r \\ \theta = 0 \to 2\pi \\ \varphi = 0 \to \pi \\ z = 0 \to L \end{cases} \qquad (27\text{-}2)$$

The line element between two proximate points in this coordinate system is:

$$ds = \alpha_r dr.\hat{r} + \alpha_\theta(rd\theta).\hat{\theta} + \alpha_\varphi(r\sin\theta d\varphi).\hat{\varphi} + \alpha_z dz.\hat{z} \qquad (27\text{-}3)$$

and the volume element is:

$$dv = r^2 \sin\theta d\theta d\varphi dr + rd\theta dz dr \qquad (27\text{-}4)$$

By calculating the space-time metric in this $[t,r,\theta,\varphi,z]$ coordinate system, they obtained:

$$ds^2 = A dt^2 - B dr^2 - C d\theta^2 - D d\varphi^2 - E dz^2 \qquad (27\text{-}5)$$

where:

$$A = 1 - \frac{2(m+q)}{r}, \quad B = \frac{1}{1 - \frac{2(m+q)}{r}},$$

$$C = r^2, \quad D = r^2 \sin^2(\theta) \quad \text{and} \quad E = \frac{z}{2(m+q)} \qquad (27\text{-}6)$$

28. Some geometrical properties of the Electrogravity field

If one tries to solve the EG field equations in a four dimensional space-time ($x_5 = 0$), and he writes the space-time metric in spherical coordinate and calculate the metric functions, then this metric will describe just the gravitational construction or gravitational curvature of the space-time. So we have:

$$ds^2 = A dt^2 - B dr^2 - C d\theta^2 - D d\varphi^2 \qquad (28\text{-}1)$$

where A,B,C and D are metric functions. Now if one solves the EG field equations in a five dimensional space-time, then the metric in 5 dimensions, will contain the construction of both gravitational and electromagnetic curvatures of the space-time.

Therefore by solving the EG field equations in 5 dimensional form, using a five dimensional metric and doing more considerations on it, it will be solved another important problem in physics, which is the geometrical nature of the electromagnetic field.

In contrast to gravity field, which it makes a curvature in the space time, in some special conditions, presence of the electromagnetic field can be visualized approximately, to makes a helical curvature in the space time with a very small radius, and from geometrical view, this can be the actual reason of the fact that, a charged particle goes on a cylindrical curved path (helical path) at the presence of the electromagnetic field.

May be someone asks that, according to the general relativity theory, when we use a cylindrical mass, it makes a cylindrical gravitational curvature around the mentioned mass, so is this curvature, the electromagnetic curvature? The answer is negative, because the radius of this cylindrical curvature is significantly bigger compared to the electromagnetic curvature's radius which is approximately in the range of 10^{-15} meter.

Of course we know that, producing this curvature by gravitational field needs a very strong   gravitational field and it is not possible even with gold metal, which has a big amount of mass density. But it is possible with some particles like electron. Even so the electron has a very small mass, but in particle theory of the electron, unfortunately until now, no one has noticed to the point that, the mass density of an electron is very big.

In classical particle theory of electron, if one calculates the mass density of an electron by dividing the mass of electron by the volume of it, (as obtained in section (2)), he will find a huge amount of mass density for it,

$$\rho_e = 9.7*10^{12} \ (\frac{kg}{m^3}) \qquad\qquad (28\text{-}2)$$

Whiles the mass density for the gold metal, is:

$$\rho_{gold} = 19320 \ (\frac{kg}{m^3}) \qquad\qquad (28\text{-}3)$$

We know that the Osmium is one of the heaviest metal elements in the nature (even heavier than gold), which the mass density of it, is:

$$\rho_{osmium} = 22.6 \ (gr/cm^\wedge 3) \ \text{ or } \ 22600 \ (kg/m^\wedge 3) \qquad\qquad (28\text{-}4)$$

and also the mass density of lead metal is: $1.134*10^4$ (kg/m^3).

As it is seen, the mass density of an electron is 9700 billion (kg/m^3) or 9.7 billion (gr/cm^3). Therefore it is very hard to find such a case in macroscopic scales, unless in some special conditions such as in the black holes.

This huge amount of mass density for electron can make a curvature with a very small radius around itself that we say, may be this is the actual source of the electron's electric field in the geometrical view of electromagnetism. This is somewhat similar to the gravitational field which is produced by a big mass in the macroscopic scales. Of course we know in non geometrical electromagnetism, the physicists believe that this electric field is related to the electric charge of an electron.

So, until here, we can conclude that the spherical curvature of the space-time with high and low intensity (small or big radius), can create two different fields. If the radius of the spherical curvature is big enough and is in the range of macroscopic scales (like the curvature around the planets and other celestial bodies), this curvature will create the gravitational field, but if the spherical curvature is very small and its radius is small enough, in a microscopic scale (in the range of microscopic particles radius, such as electron's radius), this curvature will create the electric field.

But, we must pay attention that, there isn't any stationary electron in the nature at all, because according to the uncertainty principle, if an electron comes to the rest, its momentum will be infinite, which is impossible. Therefore, there is no static electron in the nature, which only creates a static electric field. So the electron must be moving, in which case the electric and magnetic fields are produced together. Thus, from the geometrical standpoint, if this microscopic spherical curvature starts moving in a translational path with a high speed, it creates a cylindrical curvature with a very small radius, which we call electromagnetic field to it.

But there is another question here. When a Neutron, which doesn't have any electric charge (as a particle bigger than electron), starts to move, does it create electromagnetic properties around itself?

As it has been mentioned above, the electromagnetic production of an electron is related to a moving electron and there isn't any static electron in the nature, so according the E.G theory, when a Neutron accelerates, we must see some electromagnetic properties around it. Here It should be noted that, our experiments show that, some electromagnetic properties have been observed around a neutron which is accelerated.

Of course, all the experiments that have been done with neutrons up to now, have moved the neutrons at a very low speed, now if we can somehow accelerate the neutron and make it to move at a high speed, we will probably see the electromagnetic properties more clearly.

Finally, I would like to remind that, among all the microscopic particles, here we have only investigated the nature of the electric charge unit (electron) and for now, we have nothing to do with other particles and we did not investigate them. A few sentences we wrote about the properties of the neutrons are related to the answer of a physicist's question from me, that may come to anyone else's mind.

29. The actual structure of the "Electric charge unit"

Using geometrical review, I think, now it is the time to make a small correction in our imagination about the electron's structure, which is the electric charge unit.

From all forgoing discussions, one can verify that, the electron is not a particle nor wave and even no string, but it can be seen in all of these forms. Actually it is just a helical curvature in the space time with a small radius of about 10^{-15} meter, which this radius needs to be determined accurately with experiments. We have seen in quantum mechanics, it has been claimed that, an electron can be seen in both particle and wave forms depending on the observational method.

When one tries to do particle base experiments on the electron, actually he determines the radius of the helix and conclude that it is a particle with r radius, and when one tries to do wave base experiments, he determines the helix illustration on a 2 dimensional surface which is a sinus like wave, and conclude that it is a wave. The other one by determining the length of helix, conclude that it is string. Therefore, electron is nothing unless a helical curvature in the space time.

30. Electrogravity field in different scales

It should be noted that, here we have tried to look at the movement of particles from another point of view, and consider them from a novel perspective that is based on geometry. Of course, we also know that the way of looking to these issues with the old methods has also given correct answers in many cases, but there are still some issues that the old methods could not give a convincing answer to. So in general, the electro-gravitational field can be considered from three aspects:

a) In the microscopic scale, which the mass and the volume of the investigated particle are very small but the mass density is very big.

b) In the macroscopic scale, which the mass and the volume of the investigated object is very big, but its mass density is small. Like planets, stars and... .

c) In the macroscopic scale, which the mass and the volume of the investigated object is very big and its density is also very big, like black holes.

In case (a), which is about the electron, in particle theory of electron, we saw that, it creates an electric field around itself but from the geometrical perspective, we say, it creates a strong and intense spherical curvature around itself and when it performs a translational movement, it creates a cylindrical curvature with a very small cylinder radius.

Considering the intensity of this curvature, we conclude that the force corresponding to this curvature is much bigger than the force corresponding to the gravitational curvature around a planet. Therefore, the strength of this force is much bigger than the other.

We call to this type of moving cylindrical curvature with a very small radius, the electromagnetic field.

Of course, here another question that comes to mind is, whether it is possible to create an electromagnetic field with the translational movement of the gravitational curvature? For answering to this question, it should be noticed that, if the intensity of the gravitational field is very big, which means, if the radius of the gravitational curvature is very small, this work can be done. For doing this work, we need a very dense particle whose density is about 10^{13} kg/m3. The fast movement of this particle can create electromagnetic field. Of course, around a black hole where the particles that exist in it, have very big mass density and these particles are moving at a high speed along with the movement of the black hole, this case can be happen and for this reason it is predicted that, there should be a very strong magnetic field around the black holes.

In case (b), which is about the planets, due to their large mass, they create a gravitational field around them, but due to the low density, the curvature of the space time around planets is very small and is limited to the orbits around the planets. In this scale, the gravitational field seems to be the dominant field.

It seems that the structure of planets' movement around the sun in the solar system, seems to bear some resemblance to the structure of the movement of electrons around the nucleus of an atom in different atomic orbits in the particle theory of electron. When a planet which spherically curve the space time around itself, moves translational in its orbit around sun, this moving spherical curvature creates a cylindrical curvature in the space-time around it, which is corresponding to creating a field whose geometrical properties are completely similar to the properties of the electromagnetic field that an electron creates around itself with a translational motion, but its curvature radius is very big, so the intensity of this field is much more weaker than the electromagnetic field. This field needs to be further investigated, in future writings and papers for determining its properties.

In case (c), which is about the black holes and similar objects, due to their very big mass, they create a very strong gravitational field around themselves, and due to their big density, the curvature of space-time around them is very intense. When this intense curvature also performs a translational movement, the cylindrical curvature with a very small radius is created, which is the electromagnetic field.

Some geometrical behaviors of the electromagnetic field which are predicted by the Electogravity theory in this book, have been approved by some experiments like that of Kejie Fang and his colleagues[1], at the department of Physics of Stanford University.

It shows that the photons go on a circular path in presence of the electromagnetic field. This experiment approves that, the presence of the electromagnetic field makes the space time around itself, to be curved.

31. Epilogue

In this research, a unified field theory is presented for unifying electromagnetism and gravitation. As a result the Electrogravity field

---

[1] K. Fang, Z. Yu, and S. Fan, Realizing effective magnetic field for photons by controlling the phase of dynamic modulation, Nature Photonics 6, 782-787, 2012

equation is obtained. Solving the EG field equation and doing more considerations on it, can solve a significant problem in physics, which is the geometrical nature of the electromagnetic field.

In contrast to gravity field, which it makes a curvature in the space time, in some special conditions the presence of the electromagnetic field can be visualized to make another kind of curvature in the space time with a very small radius of curvature (in contrast to gravity curvature radius), and from geometrical view, for example, this can be the actual reason of the fact that, a charged particle goes on a cylindrical curved path (helical path) in presence of the electromagnetic field. This motion was justified by Lorentz force, in no geometrical thinking.

Another important result that emerged in this research is: we find out that the electric charge unit (electron) is not a particle nor wave and even no string. It is just a helical curvature in the space time. In quantum mechanics, when one tries to do particle base experiments on it, he determines just the radius of the helix and concludes that it is a particle with r radius. When one tries to do a wave base experiment, he determines the helix illustration on a 2 dimensional flat surface which is a sinus like wave, and concludes that it is a wave. The other one by determining the length of helix, concludes that it is string.

Therefore, electron is a helical curvature in the space time which its movement creates the electromagnetic field. The electromagnetic field, itself is a moving helical curvature in the space-time, which its curvature radius is very small compared to the gravitational curvature.

But about the existence of Λ_B in the field equations; it indicates the existence of a background electromagnetic field in the world, which we can talk about it more, in future writings.

Following the presentation of the Electrogravity theory, now it is the time to provide a logical answer to an important question about the beginning of the world. This stage is often more associated with the very early moments of the universe, predating the stage which is proposed for universe starting, by different models such as inflationary theory and other theories.

According to some accepted theories in this field, at the beginning of the creation of the world, when it was extremely hot and dense, the production of some particles and their anti particles such as quarks and anti quarks and proton and its anti particle started according to some special protocols and special conditions (which here we don't want to discuss about these protocols). After that, some other particles and their antiparticles were produced from the combination of these particles. Some of these particles and their antiparticles annihilated each other, releasing energy, but if all of them had annihilated each other, what happened? While the particles and their antiparticles have opposite electric charges, why are all of them not attracted to each other and turned into energy? If this had happened, our world would not have started at all.

In this case, some physicists have justified this by suggesting that, a symmetry breaking has occurred there, and the antiparticles of some particles has disappeared and leaving behind a small excess of particles, make up the matter universe we observe today. So they believe that, the antimatter part of the world has been vanished at the beginning of the universe.

However I think this reason is not logical. To facilitate understanding and answer this question, let's suppose that a proton and an antiparticle of proton have been produced in the beginning of the world, the proton has a positive charge and the antiparticle of proton has a negative charge. If the space time in which, proton and antiproton were produced, was empty, they should have recombined immediately, turning into energy and no excess of particle remains for starting the universe. But according to the E.G equations, there is Λ in equations which indicates the existence of a background Electrogravity field, which is a combination of electromagnetic and gravitational fields, $\Lambda = \Lambda_G + \Lambda_B$.

Thus, in presence of the background Electrogravity field, when the particle and its antiparticle were produced, they immediately found themselves in an electric field (the electric field component of the Electrogravity field), which causes the particles and antiparticles to accelerate in opposite directions. Due to existence of the background Electrogravity field, also an electron and a positron accelerate in the opposite directions, and in this way, at the very high temperature that existed at the beginning of the universe, the accelerating proton attracts an electron, forming a hydrogen atom, which with paying attention to the initial acceleration, it continues to move in the same direction as before, and on the other side, the negatively charged proton's antiparticle, while accelerating, attracts a positron and forms the hydrogen atom's antiparticle.

And in the same way, particles and antiparticles of other atoms are formed on both sides.

In this way, matter universe is formed on one side and the anti matter universe on the other side.

So two different universes of matter and antimatter are formed on both sides, and they moved in the opposite directions and moved away from each other.

It is important to noted that, here we have not discussed on, which theory about the beginning of the world of matter is true or false. Readers can accept any logical theory about the formation of the world of matter and its expansion which have been proposed by different scientists up to now, which the big bang and inflationary theory are two of them. The important point here is the existence of the matter and anti matter worlds, which at the beginning of the universe they separated from each other and moved in opposite directions. The aforementioned theories, big bang, inflation and ...., only talk about the events inside the matter part of the universe.

Thus we conclude that, the existence of Λ constant in equations shows

that, there must be a background Electrogravity field in all over the universe before and after the beginning of the world. Because if this field did not exist before, the matter and antimatter parts would have combined together in the first moments of the world's beginning, turning into energy again and therefore the universe would never have started.

We know that, the Electrogravity field is considered as the component of the unit field, so more generally we conclude that, before the beginning of the universe and after its beginning, there was and is a background unit field in the entire universe, that all the fields in the nature, including gravity, electromagnetism, strong and weak nuclear fields, are components of this unit field, that so far, we have been able to combine gravity and electromagnetic fields and obtain the Electrogravity equations. Also we know that, the electroweak equation already has been obtained. So, just two more steps are remaining to obtain the unit field equation. First, we need to combine the Electrogravity equations with electroweak, which does not seem to be a hard task, and then in the second step, enter the equations of the strong nuclear field into our equations and find the unit field equations.

References:

[1] A.Einstein. Letter to V. Bargmann, July 9. 1939, AE 6-207.

[2] T. Kaluza, Sitzungsber. Preuss. Akad. Wiss, 966 , 1921.

[3] A. Einstein. Äther und Relativitätstheorie. Springer, 1920.

[4] Paul Adrien Maurice Dirac, General Theory of Relativity, Princeton University Press, 1975.

[5] Subrahmanyan Chandrasekhar,The mathematical theory of blackholes, New York,

Oxford, University Press, 1983.

[6] K. Fang, Z. Yu, and S. Fan, Realizing effective magnetic field for photons by controlling the phase of dynamic modulation, Nature Photonics 6, 782-787, 2012.

[7] A. Einstein and N. Rosen. The particle problem in the General theory of relativity. Physical Review,48(1):73-77,1935.

[8] A. Einstein, The foundation of the General theory of relativity, Doc.30, 1914.

[9] Malcolm A. H. MacCallum, George F. R. Ellis and Roy Maartens, Relativistic Cosmology, Cambridge University Press, 2012.

[10] W. M. Boothby, An Introduction to Differentiable Manifolds and Riemannian Geometry, Academic Press. NewYork,1986.

[11] M. P. Hobson, G. P. Efstrathiou and A. N. Lasenby. General relativity, Cambridge University, 2006.

[12] Steven Weinberg, Gravitation and Cosmology: Principles and Applications of the General Theory of Relativity. John Wiley & Sons, New York, 1972.

[13] R. M. Wald, General Relativity, University of Chicago Press, Chicago, 1984.

[14] Davisson, C. J.; Germer, L. H. (1928). "Reflection of Electrons by a Crystal of Nickel". Proceedings of the National Academy of Sciences of the United States of America. 14 (4): 317-322

[15] Charles W.Misner, Kip S.Thorne, and John Archibald Wheeler Gravitation, 1973.

[16] Bernard F.Schutz, A first course in general relativity, Cambridge University, 2009.

[17] S. Ghayourmanesh ; E. Soufi." Electrogravity field of a charged mass object", International Journal of Science and Research, vol.12, 2023.

[18] A. Einstein, The Meaning of Relativity, 5th Ed., Including the Relativistic Theory of the Non-Symmetric Field, Princeton University Press, Princeton, 1955.

[19] S.W. Hawking and G. F. R. Ellis, The Large Scale Structure of Space-Time, Cambridge University Press, Cambridge, 1973.

www.ingramcontent.com/pod-product-compliance
Lightning Source LLC
Chambersburg PA
CBHW051416200326

41520CB00023B/7251